# The Briefest History of Time

The History of Histories of Time and the Misconstrued Association between Entropy and Time

The History of Histories of Time and the Misconstrued Association between Entropy and Time

Arieh Ben-Naim
The Hebrew University of Jerusalem, Israel

World Scientific

NEW JERSEY • LONDON • SINGAPORE • BEIJING • SHANGHAI • HONG KONG • TAIPEI • CHENNAI • TOKYO

*Published by*

World Scientific Publishing Co. Pte. Ltd.

5 Toh Tuck Link, Singapore 596224

*USA office:* 27 Warren Street, Suite 401-402, Hackensack, NJ 07601

*UK office:* 57 Shelton Street, Covent Garden, London WC2H 9HE

**Library of Congress Cataloging-in-Publication Data**
Names: Ben-Naim, Arieh, 1934–
Title: The briefest history of time : the history of histories of time and
the misconstrued association between entropy and time / Arieh Ben-Naim,
The Hebrew University of Jerusalem, Israel.
Description: New Jersey : World Scientific, 2016. |
Includes bibliographical references and index.
Identifiers: LCCN 2016000584| ISBN 9789814749848 (hardcover : alk. paper) |
ISBN 9814749842 (hardcover : alk. paper) | ISBN 9789814749855 (pbk. : alk. paper) |
ISBN 9814749850 (pbk. : alk. paper)
Subjects: LCSH: Space and time. | Time. | Entropy. | Second law of thermodynamics.
Classification: LCC QC173.59.S65 B46 2016 | DDC 530.11--dc23
LC record available at http://lccn.loc.gov/2016000584

**British Library Cataloguing-in-Publication Data**
A catalogue record for this book is available from the British Library.



Printed in Singapore

This book is dedicated to all readers of popular-science books who might reach the following logical conclusions:

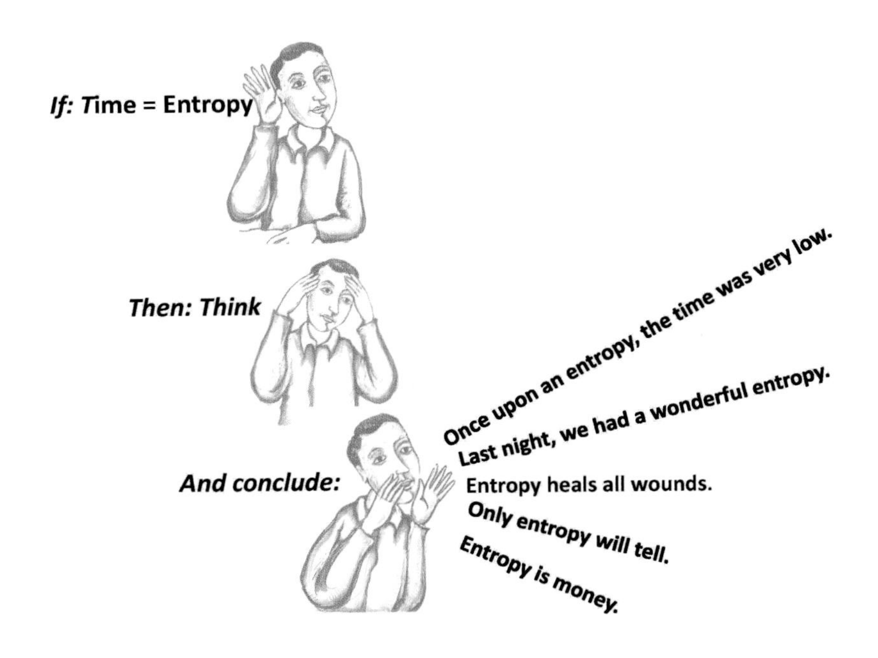


I hope that by reading this book, you will understand why all these conclusions are perfect, logical deductions from the assumption.

All the rivers run into the sea; yet the sea is not full: Unto the place from whence the rivers come, thither they return again. (Ecclesiastes 1:7)

כל-הנחלים הולכים אל-הים, והים איננו מלא;
אל-מקום, שהנחלים הולכים--שם הם שבים, ללכת
(קוהלת פרק א:ז)

# Contents

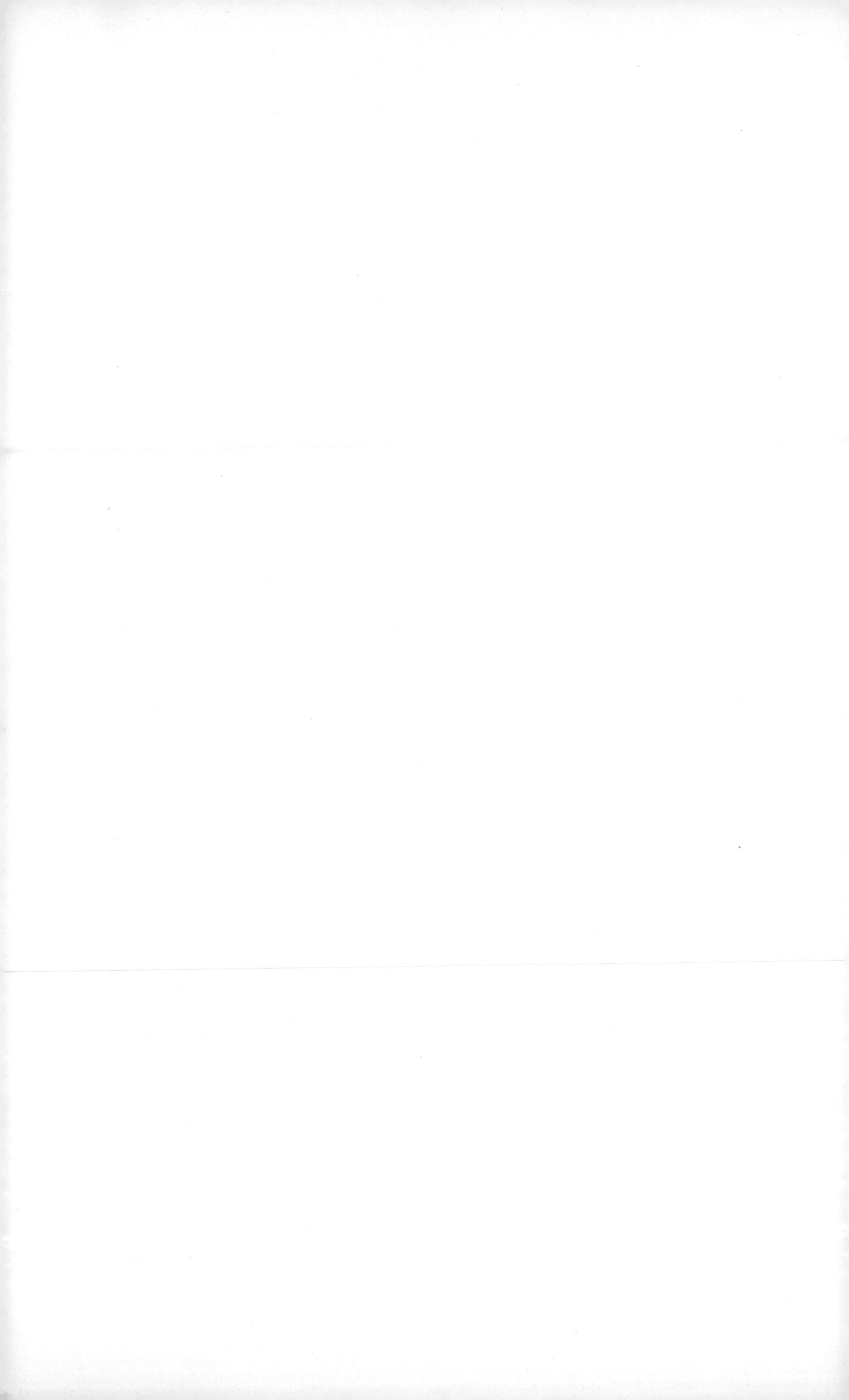

# Preface

This book is addressed to anyone who is interested in time, who is curious to know what scientists write about time, and who wants to have a *good time* reading about time. The book targets readers ranging from the layperson with no background in science to the most sophisticated scientist, all of whom will benefit from reading it.

The motivation for writing this book is threefold. First, to show that the *history of time* can be narrated in one page, or two at most; second, to examine critically what other writers write about the history of time; and, finally, to train the reader in reading scientific texts critically.

One can view this book as a sequel to Stephen Hawking's two books: *A Brief History of Time* (referred to as *Brief*), published in 1988, and *A Briefer History of Time* (referred to as the *Briefer*), published in 2005. But there is more to it than completing the sequence of *Brief, Briefer*, and *Briefest.*

As I will show in this book, the briefest history of time can be written in one or two pages. There is nothing more to say about the history of time. In fact, *Briefest* is also the *longest* history of time.

I first read Hawking's book soon after its publication in 1988. Personally, I did not like it, for the following reasons. First, its title is misleading. Most of the book is about the history of science rather than the history of time. Second, the style of the book is cumbersome, unclear, and seems like a collection of items from the history of science, from Aristotle to the very modern. A more apt title would be "From Antiquity to Modernity." What I disliked most was the misconstrued association of time with the second law of thermodynamics, which I felt should not have been written by a scientist of Hawking's stature.

In 2006, while writing my book *Entropy Demystified* (see Section 2.3), I also read the *Briefer* version of Hawking's book, published in 2005. This is certainly an improved version of the previous book. The style is clearer, and the choice of topics more appropriate. More importantly, all the parts of the first book which associate time with entropy and the second law are absent from the new version (see Section 6.2 for more details). The title of *Briefer*, however, is still misleading, as the book has very little which can be reckoned as a *history of time*.

What left me wondering was the lack of any explanation as to why certain sections of *Brief*, which deal with time, were completely removed and thus were not featured in *Briefer*. Did it dawn on the authors that those sections were wrong? Or were they omitted for compactness? I will discuss this point in more detail in Chapters 5 and 6.

This book is organized into three main parts. The first part consists of two introductory chapters: "What Is Time?" and "What Is a History of Something?". The second part consists of Chapters 3 and 4. This part is the book's core, and its content is faithful to its titles, as well as to the title of the book.

I have separated the discussion of time and space, although nowadays they are combined into space–time. Some authors write that space and time coordinates are equivalent, but this is not so.[1] The third part consists of Chapters 5 and 6. Chapter 5 discusses entropy and the second law of thermodynamics, and their relationship to time. I have exerted painstaking effort in explaining the concepts of entropy and the second law in simple and accessible language in order for any reader to grasp the message I am trying to impart. I have eliminated most of the technical details, which the reader may find in my previous books [Ben-Naim (2008a, 2012, 2015a)]. This part is essential for understanding my criticism of the books discussed in Chapter 6. Both entropy and the second law were, and are still, used in connection with the so-called arrow of time. In most cases, these concepts are misused and abused when authors write about the meaning of time. As I hope to convince the reader, there exists no connection between entropy and time. In fact, the whole of Chapter 5 has been written, somewhat paradoxically, to convince the reader that such a chapter has no place in a book on the history of time.

Entropy is not a measure of disorder, disorganization, or chaos. Entropy is not a measure of spreading of energy. Entropy is not a measure of freedom and entropy is not information and entropy is not time. If you believe any one of these "meanings" of entropy, you have to prove it!

Entropy is a special case of the Shannon measure of information (SMI). This has been proven in Ben-Naim (2008a, 2012, 2015a). The entropy of a thermodynamic system is proportional to the SMI defined on a *specific* distribution relevant to a thermodynamic system at equilibrium. In other words, while the system changes toward equilibrium its SMI changes with time. Only when the system reaches equilibrium does the SMI attain its maximal value, and this value is proportional to the entropy of the system. Unfortunately, most of what has been written about the relationship between time and entropy is a mess. This mess started with Clausius, who stated that "the entropy of the universe always increases." This is an unfortunate formulation of the second law. Clausius may be excused for this overgeneralization of the second law. But many other scientists, even today, repeat this nonsensical statement, especially when they write about the entropy of the universe at the Big Bang.

As I will argue in Chapter 6, the Big Bang is a highly speculative event. In my opinion, there are too many assumptions made in the theory which retrodict the Big Bang. None of these assumptions can be verified. Therefore, it is most likely that this event never occurred. Unfortunately, most authors writing on the Big Bang convey the impression that this is a well-established fact.

The entropy of the universe is undefinable (either theoretically or experimentally). Therefore, to say that the entropy of universe was low at the Big Bang is plain nonsense. To claim that the (meaningless) low entropy of the universe at the (highly speculative) Big Bang would explain our existence in the universe is, in my opinion, absurd, tangential to insanity.

In Chapter 6, I will also review critically both *Brief* and *Briefer*, as well as other books which discuss time, the history of time, and the theories of time. This part may be viewed as the history of the histories of time.

In conclusion, I hope to convince the reader that:

1. Time is basically *timeless*! It has no history.
2. Neither entropy nor the second law has anything to do with time.
3. The history of a substantial thing is a list of events occurring at points of space and points of time.
4. It is almost impossible to tell the history of an abstract concept such as beauty, a mathematical theorem, or time.

I hope you will enjoy reading this book.

Arieh Ben-Naim

Department of Physical Chemistry
The Hebrew University of Jerusalem
Jerusalem, Israel
Email: ariehbennaim@gmail.com
URL: ariehbennaim.com

# Acknowledgments

I am grateful to John Anderson, Andy Augousti, David Avnir, David Gmach, Robert Hanlon, Wolfgang Johannsen, Zvi Kirson, Götz Kluge, Bernard Lavenda, Azriel Levy, Xiao Liu, Mike Martin, Allen Minton, Mike Rainbolt, Eric Szabo, David Thomas, Peter Weightman, Keith Willison, and Harry Xenias, for reading parts of or the entire manuscript and offering useful comments.

Special thanks to my friend Alex Vaisman, for the beautiful illustrations in the book.

As always, I am very grateful for the help I got from my wife, Ruby, and for her unwavering involvement in every stage of the writing, typing, editing, re-editing, and polishing of the book.

# List of Abbreviations

| | |
|---|---|
| 20Q | 20 questions |
| BH | black hole |
| *Brief* | *A Brief History of Time* |
| *Briefer* | *A Briefer History of Time* |
| *Eternity* | *From Eternity to Here* |
| *Begin and End* | *Did Time Begin? Will Time End?* |
| second law | second law of thermodynamics |
| SMI | Shannon measure of information |

# 1

# What Is Time?

## 1.1 Introduction

I would like to begin by addressing the question posed in the title of this chapter to you, the reader of this book. I am sure you know what *time* is; often you use this word when asking for the time or telling the time. I suggest that you take a few moments (of time, of course) to do a little homework, or an exercise. Normally, you do an exercise *after* you learn or read something new in order to assess your understanding of what you have learned. Here, I suggest that you do the exercise in order to prepare yourself for critically examining what you will read in the following pages of this book.

I suggest that you take a look at Table 1.1, where you will find a list of idiomatic expressions, all involving time. More, with some illustrations, appear in the Appendix.

**Table 1.1 Idiomatic Expressions Involving Time**

*Quality of time:*

- Good time
- Bad time
- Hard time
- Have a nice time
- Have a good time
- Time is ripe
- Have a rough time
- Have a thin time
- Quality time
- Big time
- Time is swift
- A devil of a time
- Have an easy time of it
- This was the time of my life
- Have a lovely time
- Take your time

*Quantity of time:*

- Short time
- All the time in the world
- Long time
- Pressed for time
- Run out of time

*Value of time:*

- Time is money
- Don't waste your time
- Buy time
- Save time
- Invest time in
- Live on borrowed time
- Lose time
- A waste of time

(*Continued*)

**Table 1.1 *(Continued)***

*The notion of motion and speed associated with time:*

- Time passed
- A race against time
- Arrow of time
- Time runs fast
- Time stopped moving
- Time runs slow
- Beat time
- Time flies
- Time flies when you are having fun

*Personification of time:*

- Time will tell
- Time heals
- Time ravages
- Time works wonders
- Kill time

Please read the list slowly and carefully. Write down the meaning of each expression, then ask yourself several questions, and try to answer them. Examples are:

1. Do all the "times" in the list refer to the same time?
2. Are all the "times" in the list the same as the time you *feel* passing by?
3. Are all the "times" in the list the same as the time you *read* on your clock?
4. Why does time have so many attributes, such as "good" and "bad," "fast" and "slow," and many more?
5. Can you think of other abstract concepts, such as space, love, or beauty, to which so many attributes have been imparted?

6. Why do all the idioms on the list sound "natural" but if you replace "time" by "space" in any of them you will get an awkward and perhaps meaningless expression?
7. Finally, ask yourself: What is time?

I suggest that you try to answer these questions in writing, or mentally, before you continue to read this book. Perhaps you can come up with some more questions and answers, and if you feel that these can help future readers of the book in understanding time, please write to me and I will add them to the list of questions I suggested above.

## 1.2 Can We Define Time?

Before we start our journey in time and about time, I would like to introduce the following notation. Most of the time, I will use "time" in its colloquial sense, such as all the "times" appearing in Table 1.1. However, sometimes, when I want to emphasize that I am referring to the *real, physical* time, I will use the notation "Time." For those who might question the existence of Time, let me clarify what I mean by "physical time." This is the Time I read on my clock. This time may be different from the time you read on your clock, and the times read (or not read) on other clocks around the world. It is not possible (perhaps even meaningless) to talk about the *same* Time for everyone who reads the Time on his or her clock. We usually measure *differences* in Time in an experiment, for example when we want to determine the speed of an object. A delightful book on the various ways of measuring Time is the one by of Hooft and Vandoren (2014).

Everyone who is asked the question "What is time?" knows what time is, and yet a precise definition of time is as elusive as definitions of so many abstract concepts, such as space, beauty, or life. [Carroll (2010) in his book From Eternity to Here promises in the prologue: "By the end of this book we will have defined *time* precisely, in ways applicable to all fields." As we will discuss in Chapter 6 such a promise was never fulfilled].

A famous quotation by St. Augustine goes: "What then is time? If no one asks me, I know what it is. If I wish to explain it to him who asks, I know not." We use the concept of time more than once almost every day, without really bothering as much about its precise definition, or whether such a definition exists at all. In fact, we use or come across "time" in numerous idiomatic expressions, which have very different meanings than physical Time.

The periods of Time which we may have experienced in our lives at one time or another, like "a bad time," "a good time," or even "a marvelous time," are not measured by our clocks. Does time really have qualities attached to it, qualities which change from Time to Time, and from person to person?

Does time have value? We certainly believe that it has when we say "Time is money" or "Time is gold," when we "buy time," "invest in time," or try to "save time," or are careful not to "waste time."

Why do we personify Time and impart to it several personal attributes? Does time really heal all wounds? Does it ravage anything? We say "Time works wonders" or "Only time will tell," but does Time really work or tell anything? And while we are still in the process of assigning personal attributes to time, will we then want to "kill time," because it is "so bad"?

I do not know of any other abstract concept which has so many attributes attached to it. In modern physics, "Time" has received almost the same status (some say even equivalent) as *space* — more precisely, as the three coordinates ($x$, $y$, $z$) we use to define a point in space.

Can we say that *space* runs fast or slow? Can we say that *space* is bad or good, or wonderful? Sometimes we do say "buying space" or "saving space." This is very different from saying "buying time" or "saving time." When we say buying space, we mean the real *physical space* in which we place some objects. These two expressions are the exceptions; most of the idiomatic phrases in the table cannot be used with "space" instead of "time." For instance, we never say that "space heals," or that we want to "kill space."

I cannot explain why so many attributes of time seem so "natural," while using the very same attributes assigned to space will sound awkward. Perhaps *time* does not have the same status in daily lives as *space*. We feel that time is more important to our lives than space. This is an illusion, for without space there will be no place for us to be. Yet, we have a sense that since time is given to us in a "limited amount" we should maximize and use it as best we can.

Smolin (2014) has devoted almost an entire book to the question of whether time is only an illusion or is real. Personally, I do not think that this question is answerable within physics. This is a philosophical question similar to the question about the reality of everything we perceive with our senses, including our very own existence.

In the next sections, we turn to some aspects of time which are discussed by authors of popular-science books.

## 1.3 Does Time Flow?

Perhaps the most dominant feeling we have regarding time is that it flows, and it runs, sometimes as quick as a flash, sometimes excruciatingly slow, but it moves nevertheless. But, does Time really flow?

Before we try to discuss the question of *time flow*, However, consider the following examples:

In an encyclopedia you find sentences like:

The Jordan River flows into the Dead Sea.
The Mississippi River flows into the Gulf of Mexico.

Everyone understands that the "flow of a river" is a figure of speech. It is not the river that flows, but the water. Indeed, in an encyclopedia you might find the more accurate expression, about the Jordan River draining into the Dead Sea. It is not the river itself that flows but rather what is in the river.

And in the Bible we find these words (see dedication page):

All the rivers run into the sea; yet the sea is not full;

Of course, there is no reason for the sea to get full (of water), if only the *rivers* run into it....

When we say something flows, we need to attach some *speed* and *direction* to the flow — say, a canoe floating on the river might be traveling at 10 meters per second, in the direction toward the Dead Sea. The same applies to the water which flows in the Jordan River. Also, we can say that the water in the river drains into the Dead Sea, at a rate of say, one cubic meter of water per second.

Clearly, the river itself (normally) does not flow; it has no speed, no direction and no arrow.

Of course, it is meaningful to say that *the river flows*, in which case the river in its entirety (or part of it) is moving in some direction and at a certain speed. The point to emphasize is that the speed and the direction of motion of the river itself are not along the river, but along some other arrow. Figure 1.1 illustrates the difference between the water flow and the river flow.

Imagine that you are riding in a canoe along the Jordan River. You feel that you are moving relative to the riverbanks. You can tell the speed and direction of the motion of the

**Fig. 1.1.** The water in the river flows into the sea.

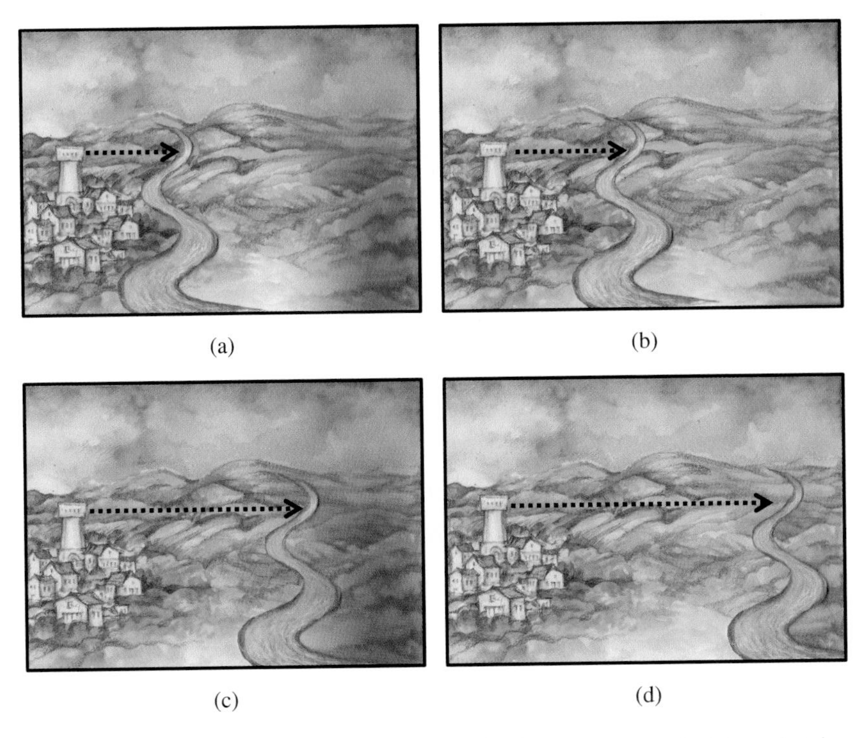

(a) (b) (c) (d)

**Fig. 1.2.** The river flows relative to the house on the left hand side.

canoe relative to the riverbanks, but you cannot tell whether the *river itself* is moving or not unless you look at its motion relative to a point outside the river. Only then can you tell the speed and direction of the motion of the river itself. Figure 1.2 shows how the river itself moves relative to a nearby farm.

When we talk about Time's flow, we say that time runs by, time runs fast or slow, but we do not mean that time itself is running. *Events* change, or flow, or unfold *in time*. Thus, the river is the analog of time, and the water is the analog of events. The clock ticks about sixty times per minute; this is the speed of the ticking. The earth rotates about its axis once a day (that is how we define a day). It revolves around the sun once a year (that is how we define a year). We can also define the rotation of the earth in degrees per second, or any other units. All of these events are *registered* or *recorded* on the *time scale*, the *time axis*, or the *time line*. If we want to talk about the flow of Time itself, then we have a problem. First, we do not know whether it moves — and if it moves, does it move in space, like the river itself moving from one location to another? And what is the speed of the motion of Time? We cannot register the points of Time in which it passed on the same Time scale or Time axis. This will be the same as recording the sequence of locations of the *river as it moves along the river* (as we do for the locations of water moving along the river). Thus, if we want to register the flow of Time, we cannot do this on the same Time axis. We need to imagine another time axis — let us call it *super-time*, S-time — on which we record the points of S-time along which Time flows. This would also be true of any other event that we might imagine Time has undergone, or gone through, not on the Time axis itself, but on the S-time axis. This will be true of the record of all the history of Time (see Chapter 4), including the beginning and the end of time… .

But if we create a new super-time axis we might as well ask whether the S-time flows. Again, it cannot flow along points of S-times of itself, and we will be forced to record the points of S-times on yet another super-super-time axis, which we might denote by S-S-Time, and we can continue to do this forever. This is shown schematically in Figure 1.3.

To avoid all these fantasies, we better admit we do not know whether time flows or not. We do not know whether time changes (where and when?), and we do not know whether time has or has not a history (see Chapter 4).

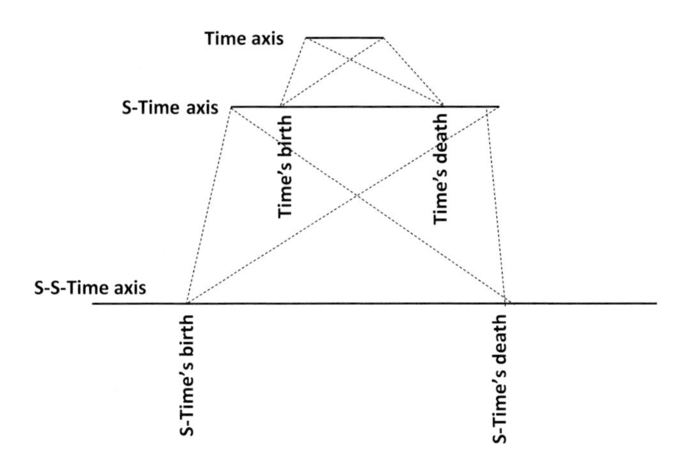


**Fig. 1.3.** Time axis, S-Time axis, and S-S-Time axis.

Thus, the answer to the question "Does Time flow?" is very simple. Whatever your answer is, it cannot be verified and it cannot be refuted. It is up to you to choose.

## 1.4 Is There an Arrow of Time?

Everyone "feels" that time always goes in one direction — it always increases. The hands of a clock always turn clockwise. The next time you celebrate your birthday, the day number and the month number will be the *same*, but the year number will change — it will always be a *larger* number.

In some popular-science books, the "arrow of time" is defined as the arrow pointing from the past, through the present, and to the future. But this definition is not less problematic than the statement that time flows in the direction of the arrow of time. Can we define an arrow of time which is objective, and will be accepted by everyone? We will discuss a few more specific arrows of time in the next

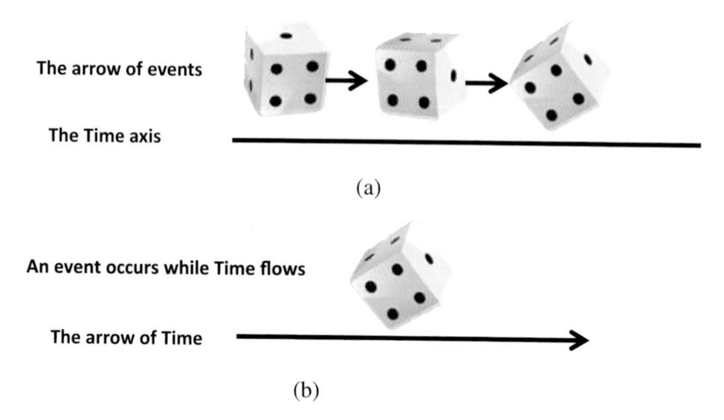


**Fig. 1.4.** Two views of the "flow of time": the arrow of events and the arrow of time.

sections, but here we ask the more basic question about the arrow or the direction of time.

In my opinion, there are two possibilities we can choose from. We can either assume that Time is a coordinate, an abstract line, not moving, not flowing, and not having a preferred direction, and *we* or any *event* is "moving" along this line in a preferred direction; see Figure 1.4a. Here, "moving" is not in space, but *in time*. Or, we can assume that while any event is occurring, "time moves" in its own preferred direction, which we call the arrow of time; See Figure 1.4b.

It is impossible to decide which one is true. I am in favor of the first view, but both could well be illusions.

Sometimes, authors of popular-science books argue that the "fact" that we can influence the *future*, but not the *past*, is connected with the Second Law.

For instance, Deutsch (1997) writes:

> ... we think of the flow of time in connection with causes and effects. We think of causes as preceding their effects.

> The common-sense view is that we have *free will*: that we are sometimes in a position to affect future events (such as the motion of our bodies) in any one of several possible ways, and to choose which shall occur; whereas, in contrast, we are never in a position to affect the past at all.

I believe that what the author means by "common-sense view" is our common subjective experience. I certainly share this experience that with our "free will"; we can affect the future, and not the past. I doubt however, the physical reality of this subjective experience. The fact that we *feel* that we can affect the future might be an illusion since we can never predict the future, and we can never tell whether any event was, or was not, affected by our "free will." For instance, I decide now to write the following sentence, and I observe that it is written as I have "predicted." It is not clear that my "free will" was the cause of the event. Perhaps my own "free will" was determined by some other cause, say some biochemical process in my brain which caused me to decide to write a sentence. Clearly, we are biased by the events that do occur in the future which are consistent with our "free will," but we can never be sure that the occurrence of these events is really a *result* of our own decision. In fact, we cannot be sure that our decision to do something was not predetermined, i.e. our free will itself, or even the *feeling* that we have a free will, might be an illusion.

In any case, the subjective feeling of affecting the events in the future has nothing to do with the second law; as so many authors claim.

There is another common confusion between cause and effect, on one hand, and conditional probability, on the other hand. We will not discuss this subject in this book. Some examples are discussed in Ben-Naim (2008a, b).

In the following sections, we will discuss the three most common "arrows of time." Some authors define five or more arrows. Newton (2000), p. 150, defines the following arrows:

1. The delay between cause and effect.
2. The psychological awareness of the passage of time, pointing toward the future — some call this the biological arrow. I will call it *cognitive*.
3. The unidirectional flow of time embodied in the second law.
4. The cosmological arrow, defined by the expansion of the universe.
5. The direction of the time parameter used in physics, the directed fourth dimension of space–time. . . .

In my opinion, three arrows is already too many. I could invent many other arrows, which do not contribute anything to our understanding of Time. For instance: The number of books published in the world always increases with time. Can we refer to this as "the book's arrow of Time"? Yes, we can, but the question is whether such a definition is useful in understanding time.

## 1.5 Is There a Psychological Arrow of Time?

Of course there is! We *feel* that time flows in one direction. Whether it runs fast or slow, it is always in the same direction. We never experience time flowing "backwards." So there is a definite *sense* of direction — time always increases, and that is almost by definition the psychological arrow of time. Not only is there a sense of direction, but there is also a sense of *speed* for the time. On one hand, we sometimes feel that time runs fast, especially when we are busy or having a ball. On the other hand, we feel that time runs excruciatingly slow when we are in pain or are suffering, or when eagerly waiting for an important message to arrive. There are times when time comes to a halt, or stops flowing perhaps when we are in slumber or when we pass out.

All these are valid experiences. The only question begging to be asked is whether this arrow of Time has any physical reality. Can we quantify it, is it measurable? The answer is "Probably not." It is unfortunate that some people identify this psychological time arrow with either the thermodynamic arrow or the cosmological arrow (see Chapter 6).

Some authors invoke the fact that we remember the past but not the future, as the physical basis of the psychological arrow of time. It is even suggested that this fact bridges the divide between the psychological arrow of time, which is mainly subjective, and the thermodynamic arrow of time, which is supposed to be objective (see Section 1.6). Not only is such a bridge presumed to exist, but some people conclude that since the psychological arrow of time is subjective the thermodynamic arrow must be subjective too. Others reach the opposite conclusion: Since the thermodynamic arrow must be objective, it follows that the psychological arrow must be objective too. Both of these inferences do not make any sense. There is no connection between the subjective psychological arrow of Time — the arrow which all of us feel, and which cannot be denied — and the fictitious (or perhaps I should say, the illusionary) thermodynamic Arrow of Time (see Section 1.6).

Thus, the existence of the psychological arrow of time is independent of whether or not an arrow of time exists. We *feel* that time moves in one direction — and therefore this is the definition of the psychological Arrow of Time.

Our ability to estimate the Time it takes, to do something is different from the psychological sense of time. Here, we rely on our experience in guiding us as to how to estimate the time it takes to walk, or to drive from one place to another. I believe that most animals have some kind of sense of time. A falcon detecting a dove at an extremely large distant point must be able to "calculate" the time it will take to reach the dove. However, if the falcon wants to intercept the dove, while the latter is flying, it has to estimate how far the dove will be moving during that period of time, say from A to B; see Figure 1.5. This kind of sense of time is probably a trait of most predators which must hunt to survive. Of course, the same is true of the prey which must evade being hunted!

## 1.6 Is There a Thermodynamic Arrow of Time?

The idea of a thermodynamic arrow of Time is quite old, and probably started with Clausius' formulation of the second law: "Entropy of the Universe always increases."

It is based on the frequently made statement that *entropy* always increases, and since time also increases, one assumes that there must be a correlation between the two; hence the existence of a thermodynamic arrow of time. The actual connection between thermodynamics and the arrow of time is probably due to Eddington (1928). Incredible as it may sound, some even *equate* (with an equality sign!) entropy with time.

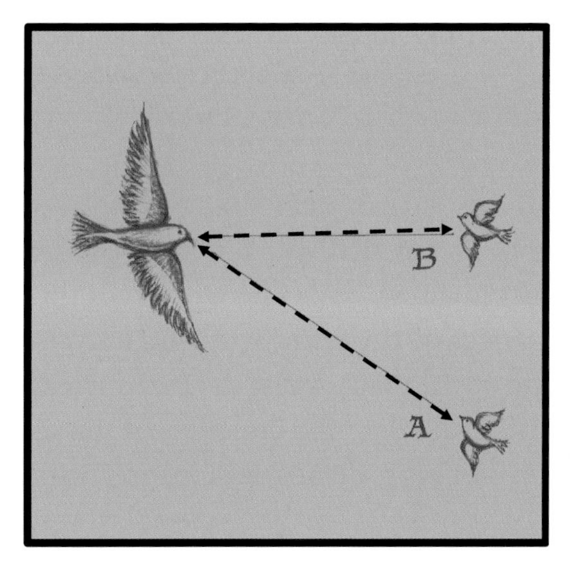


**Fig. 1.5.** The falcon has to estimate the time it will take to get to the dove at A. If the dove is moving, then it has to estimate the distance it will travel and how long it will need to get to B.

"Entropy not only explains the arrow of time, it also explains its existence; it is 'time.'" [Scully (2007)]; see also Chapter 6.

Such fantastic ideas are the result of a profound misunderstanding of what entropy means, and what the second law states.

First, entropy in itself does not increase or decrease. It is as meaningless to say that "entropy increases" as to say that "beauty increases."

One must specify the system for which the entropy either increases or decreases, as we specify the person whose beauty is enhanced or diminished.

What people usually mean is that the entropy of the universe always increases. For example, Atkins (2007) writes:

> The entropy of the universe increases in the course of any spontaneous change. The keyword here is *universe*; as always in thermodynamics, the system together with its surroundings.

Unfortunately, such statements are empty. The entropy of the universe is not defined, much as the beauty of the universe is not defined. It is therefore meaningless to talk about the changes of the entropy in the universe. Yet, most writers on entropy do talk about the ever increasing entropy of the universe, and relate the direction of this change to the arrow of time. [For more details, see Chapter 6 and Ben-Naim (2015).]

The entropy is defined for a well-defined thermodynamic system. This means that if you have a glass of water or a bottle of wine at some specific temperature $T$, pressure $P$, and some composition $N$ (and neglecting all kinds of external fields, such as electric, magnetic, or gravitational), then the entropy of the system is well-defined. It is a function of the specified variables; we write it as $S = S(T, P, N)$, and we say that the entropy is a *state* function. This means that for each *state* of a thermodynamic system there is a corresponding value of the entropy. For some simple systems, such as ideal gases, we can calculate the *value* of the entropy. For some more complicated systems, we can measure the entropy up to an additive constant or, equivalently, we can measure the changes in entropy for the same system at two different states. We write it as $\Delta S = S(T_2, P_2, N_2) - S(T_1, P_1, N_1)$. In this characterization of the system the entropy is a function of the variable $T$, $P$, $N$. It *is not* a function of time, either explicitly, or implicitly through the variables $T, P, N$ which are unchanged for each equilibrium state.

There are special sets of variables characterizing what is called an *isolated system*. Such a system is characterized by a fixed energy $E$, fixed volume $V$, and fixed composition $N$. Again, we neglect any effect of an external field on this system. Clearly, such a perfect isolated system does not exist. There are always some interactions with external fields (such as gravitational) which affect the state of the system. However, as an idealized case it is a very convenient system, and it is also the system on which the entire edifice of statistical mechanics was erected.

Given a *state* of an isolated system ($E$, $V$, $N$), its entropy is defined. Again, the entropy of this system is a function of the variables $E$, $V$, $N$. It is not a function of time.

Here is where the time sneaks into thermodynamics. The second law states that if we remove any internal constraint of an isolated system the ensuing spontaneous process will always cause entropy to either increase or remain unchanged. The simplest example is the removal of the partition separating two different gases, as shown in Figure 1.6. Once the partition is removed, the system will proceed to a new state (shown on the right hand side of the figure), having higher entropy.

The second law does not state that *entropy* increases with time. It does not state that the entropy of *any* system increases with time. The second law does not state

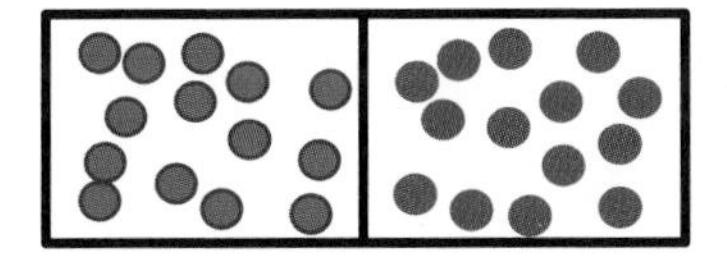
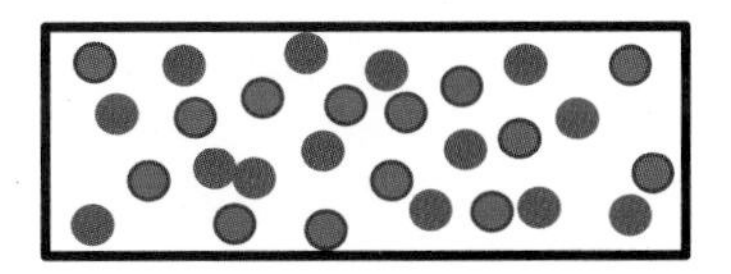

**Fig. 1.6.** Spontaneous mixing of two gases.

(although some say so) that the entropy of the universe always increases. It does not even state that the entropy of an *isolated* system increases with time. Entropy is simply not a function of time. Hence, there is no thermodynamic arrow of time!

What the second law states "is that when we remove a constraint of an isolated system, the system will move to a new equilibrium state, having higher entropy." Thus, $\Delta S > 0$ is the difference in the entropy of a given system at *two different states* — not at two *different times*. It is true that the positive change of entropy always occurs in a positive change of time. This does not mean that the entropy is a *function* of time. We can use the metaphor of a ball rolling down the hill; see Figure 1.7. Suppose a ball is placed at a certain location on the hill, having height $h_1$. We release a constraint, and as a result the ball falls to a new height, $h_2$. The time it takes for the ball to roll down the hill depends on the size of the ball, the friction with the earth, and the air, etc. For different experiments, there will be different sequences of heights. One cannot say that the height of the ball is a *function* of time, or a decreasing function of time. (Note that the "driving force"

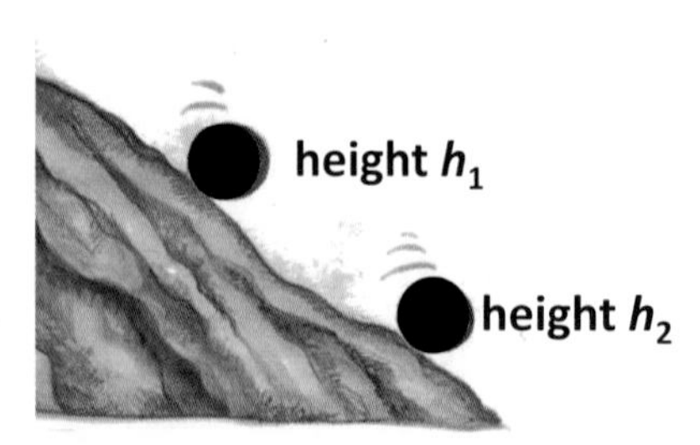


Fig. 1.7. A ball rolling along the slope of a hill.

for the motion of the ball is very different from the "driving force" for the increase in the entropy.) Similarly, the number of letters I am writing increases with time. This does not mean that the number of letters is a function of time. We will further discuss this question in more detail in Chapter 5.

## 1.7 Is There a Cosmological Arrow of Time?

While the psychological arrow of Time is a subjective feeling we have, and the thermodynamic arrow of time is a result of the misconception about entropy and the second law, the cosmological Arrow of Time is neither felt by us nor pertains to cosmology. It is pure nonsense.

In many popular-science books (see Hawking's book *Brief*, discussed in Chapter 6), a cosmological arrow of time is defined as "the direction of time in which the universe is *expanding* rather than *contracting*." [Hawking (1988), p. 145]. This is clearly meaningless. It becomes even senseless when people infer that if and when the universe *contracts*, the direction of time will be *reversed* (whatever this might mean).

We do not know whether or not the universe will ever stop expanding or start contracting. Whatever the fate of the universe will be, it has nothing to do with the time arrow we feel, or with the time arrow presumed to be associated with entropy. Whatever the time is, and whatever the arrow of time is, it is not determined by either the expansion or the (possible) contraction of the universe. We will further discuss this arrow of time in connection with Hawking's book Brief in Chapter 6. Fortunately, all these discussions in Brief about the cosmological arrow of time were removed in Briefer — and for a very good reason!

## 1.8 Does Time Have a Beginning, and Will It Have an End?

Among all the questions associated with Time, these two have received the most attention. They feature in most popular-science books, and even entire books are devoted to them (see Chapter 6). Why? As some authors claim, answering these two questions will have a *profound* effect on our life, will dramatically change the way we view our place in the universe and, of course, will answer the most fundamental questions regarding life; Where did we come from, and where will we be going at the end of time?

Such baloney fills pages upon pages of books which justify the extended discussions of these questions (see also Chapter 6). Personally, I do not think any of the answers to these questions (if there will ever be such answers) will have any noticeable impact on our lives, our view of our place in the universe, and so on. In fact, I doubt that these questions occupy the minds of even a tiny fraction of humankind — and I think probably none of the other living creatures.

I can think of many more interesting and important questions regarding time, the answers to which will have a far more profound impact on our lives: Can we *control* the rate of time changes? Can we use time more efficiently? Can we *save* time, and *buy* time (not in their idiomatic sense, but really saving and buying)? Can we choose, or design the quality of time?

All these questions sound fanciful, and it may well be that they are indeed. But the same is true of the two questions posed in the title of this section (as well as in the title of at least one book; see Section 6.4).

The point I wish to make here is that whatever the reality of the questions I raised above may be, if we find an answer to even one of them, it will have the most profound effect on our lives — far more than the answers regarding the beginning or the end of the universe.

The reason why cosmologists talk so much about these two particular questions is that they have a theory of cosmology, or they hope to have a theory of the universe. In such a theory, Time and space are parts of the universe. Therefore, any theory of the universe will also be a theory of Time and space. If the theory can predict the fate of the universe, and retrodict the origin of the universe, then we will have answers to these "central questions in our lives."

Unfortunately, authors who write about the theories of the universe do not emphasize strongly enough the validity (or invalidity) and the limitations of these theories.

Suppose we have a theory which correctly describes the current expansion of the universe. Now, we apply this theory to predict the evolution of the universe for the following billions of years, or extrapolate back to retrodict the state of the universe billions of years ago. Unfortunately, we can never be sure that the theory's predictions are valid for such a huge span of time, nor can we trust the validity of the theory itself for situations which might be (or have been) so extreme at such distant times.

From what we observe today, we conclude that the universe is expanding (see Chapter 6 for more details). One can, in principle, extrapolate back in time and find that some 13.7 billion years ago the entire universe was concentrated at a singular point having infinite density (and perhaps infinite temperature; see Chapter 6). This is called the Big

Bang. At the Big Bang, Time and Space were born! Some authors tell you that it is *meaningless* to ask what happened *before* the Big Bang. In my view, it is far more *meaningful* to ask such a question than to claim that it is *meaningless* to do so.

Similarly, various theories predict that billions of years from now the universe will shrink to a singular point. That would be the end of Time (and space). This is called the Big Crunch. Some authors tell us that it is *meaningless* to ask what will happen after the Big Crunch. Again, in my view, it is far more *meaningful* to ask about what will happen after the Big Crunch than to claim that it is *meaningless* to do so.

Some authors do indicate that we still do not have a complete and reliable theory of the universe, and even if we did, it would not be clear whether or not all the laws of physics which are part of this theory would be valid under the extreme conditions, predicted to be at the Big Bang (or the Big Crunch). In my view, we will never be able to trust any theory to be valid billions of years from now, either into the past or into the future. Therefore, I do not believe that the questions posed in the title of this section are answerable. Furthermore, any tentative or speculative answers given to these questions would have no effect on our lives, or how we view our place in the universe. More on this in Chapter 6.

## 1.9 Can We Travel into the Past or into the Future?

Traveling into the past or into the future is a favorite topic of science-fiction books and movies. Some popular science books also discuss these questions, with varying degrees of seriousness.

Those who discuss traveling in the past usually claim that such travel leads to a paradox. Imagine that you go back, say a hundred years, and you meet your grandmother when she was a young girl, long before she tied the knot. You have an argument with her and kill her. Clearly, as a result of her untimely death, she would not be able to marry, your mother would not have been born and, of course, you would probably not have been born either. Is this a paradox? No, it is not a paradox at all. A paradox is an unacceptable conclusion, or an unreasonable result based on a reasonable assumption or an acceptable premise [Sainsbury (2009)]. In the case of traveling into the past, one cannot claim that the very travel is a reasonable assumption, or an acceptable premise.

Besides, if you could really travel into the past, say the year 1900, when you were not born yet, and you meet your grandmother ... but who exactly meets your grandmother at this date since you still did not exist, so you cannot *be* there to meet anyone.

If you believe that you, in the flesh as you are today, can be present in the year 1900, and kill your grandmother, then what? You come back to the present to find that you were not even born, so you cannot enter the time machine, cannot meet your grandmother, and cannot kill her.

Similarly, there are all kinds of fanciful stories about traveling to the future, and getting into paradoxes, no less puzzling than paradoxes involving traveling into the past. In my opinion, traveling into the future is both absurd and trivial.

It is absurd because the future does not exist. It is also not clear at all whose *future* you will be visiting. What does it mean that you will travel to, and *be* in a future, when you will, with a high probability, be dead? Will you suddenly appear in a world where your grandchildren are living? Will you join them in visiting your own grave? Besides, on that particular visit when you are dead, would you be able to enter the time machine to travel back into the present? If you believe in parallel universes (I don't), then you can visit all possible futures. That must be fun!

It is also trivial to travel into the future. You do not need any time machine. You are *traveling* into the future every minute of your life....

All these stories and paradoxes have no place in science. Their rightful place is in science-fiction books.

## 1.10 Does Time *Ravage* Anything?

Some authors like to use the phrase "the ravages of time and entropy." [See Ben-Naim (2015b).] We will discuss the more frequently used phrase "the ravages of entropy" in Section 5.10. Here, we focus only on the *ravages of time*. What is meant by this expression is that things tend to be destroyed, decay, and die (Figure 1.8). In particular, when people associate (quite frequently) the arrow of time with the arrow of entropy, and in addition associate entropy with disorder (see Chapter 5), they almost universally conclude that as time goes on, order turns into disorder, structures turn into rubble, life into death — and eventually "predict" that in the long run the whole universe is doomed to

**Fig. 1.8.** Ravages of time and entropy.

death (or thermal death). At that time disorder will prevail through the entire universe — and that will also be the end of Time!

The ultimate fate of the universe "is not ours to see." Borrowing a few lines from the famous song: "*Que sera, sera*, whatever will be, will be...the future's not ours to see, *que sera, sera*." Therefore, the ultimate fate of the universe cannot be predicted by any sound, or any respectable theory. We will discuss this topic in Chapter 5. For now, let me draw the attention of the reader to the following facts:

Ever since life began on this planet, life forms, variety and complexity have only *increased*. Life certainly cannot be viewed as a result of the ravages of time. You can say that someone died as a result of the ravages of time, but this is only a figure of speech applicable to this particular event. You can also say that every birth, every building, or every work of art is the result of the *constructive* power of time. This too is a figure of speech, no less accurate from "the ravages of time." The truth is that time does not ravage anything, does not construct anything, does not give birth, does not snuff out anyone's life. Simply put, Time does not do anything. This is *a fortiori* true for entropy. (See also Section 5.10, on the ravages of entropy.)

## 1.11 A Time for Everything

In this chapter, we ask many questions. Some have answers, some others do not, and still others may never have an answer. It is Time to switch from questions to assertions. Read Table 1.2, and take out of it whatever you wish. There is a Time for everything, and there is a time to end this chapter,

**Table 1.2 There is a Time for Everything**

There is a time for everything, and a season for every activity under the heavens:

a time to be born and a time to die,
a time to plant and a time to uproot,

a time to kill and a time to heal,
a time to tear down and a time to build,

a time to weep and a time to laugh,
a time to mourn and a time to dance,

a time to scatter stones and a time to gather them,
a time to embrace and a time to refrain from embracing,

a time to search and a time to give up,
a time to keep and a time to throw away,

a time to tear and a time to mend,
a time to be silent and a time to speak,

a time to love and a time to hate,
a time for war and a time for peace.

— Ecclesiastes. 3

and a time to start a new one. It is time to relax, and enjoy the verses from "Ecclesiastes."

לַכֹּל, זְמָן; וְעֵת לְכָל-חֵפֶץ, תַּחַת הַשָּׁמָיִם. (קהלת פרק ג)

עֵת לָלֶדֶת, וְעֵת לָמוּת
עֵת לָטַעַת, וְעֵת לַעֲקוֹר נָטוּעַ

עֵת לַהֲרוֹג וְעֵת לִרְפּוֹא
עֵת לִפְרוֹץ וְעֵת לִבְנוֹת

עֵת לִבְכּוֹת וְעֵת לִשְׂחוֹק
עֵת סְפוֹד וְעֵת רְקוֹד

עֵת לְהַשְׁלִיךְ אֲבָנִים, וְעֵת כְּנוֹס אֲבָנִים
עֵת לַחֲבוֹק, וְעֵת לִרְחֹק מֵחַבֵּק
עֵת לְבַקֵּשׁ וְעֵת לְאַבֵּד
עֵת לִשְׁמוֹר וְעֵת לְהַשְׁלִיךְ
עֵת לִקְרוֹעַ וְעֵת לִתְפּוֹר
עֵת לַחֲשׁוֹת וְעֵת לְדַבֵּר

עֵת לֶאֱהֹב וְעֵת לִשְׂנֹא
עֵת מִלְחָמָה וְעֵת שָׁלוֹם.

Note that "time" appears in the heading and in the rest of the text. In the Hebrew version, the word "time" appears in two versions. The first means time, and the second means "Time appropriate for… ."

**Does the Time change or the clock?**

**An old sun clock in Jerusalem. Picture taken at 11:12 am**

# 2

# What is a History of Something?

The purpose of this chapter is to prepare you, the reader, for reading critically books which tell you the history of Time.

In this chapter, we present eight histories — brief, or rather very brief, histories of several things. As we will see, there is hardly any difficulty in narrating the history of something substantial. As examples, we will present a brief history of mankind, of me, of a specific book I wrote, and of a specific object. In all of these, history can be defined as a sequence of events associated with a particular object or a person which occurred at some particular points in Time (or a period of time), and at some particular points in space (or a region in space).

When we want to tell the history of an abstract concept, chances are that sometimes there is a blur or fuzziness between two different histories: the history of the concept itself, and the history of ideas, perceptions, interpretations, etc. related to the concept. We will discuss a few of such histories of concepts; beauty, a mathematical theorem, a law of physics, and the theory of evolution.

Having an idea about the history of an object, and that of an abstract concept, we turn to the history of *space*, before discussing the history of Time. Time and space are considered as part and parcel of our reality. They are *real* in the sense that we carry out experiments *in space* and *in Time*. However, they are abstract in the sense that we cannot carry experiments *on* either *space* or *Time*. In other words, we can narrate events which occurred *in Time* and *in space*, but we can hardly narrate events which occurred *to Time* or *to space*, except perhaps their beginning and their end. We will discuss this difficulty in Chapters 3 and 4.

## 2.1 A Brief History of Mankind

It is difficult to pinpoint either the date or the location of the "dawning" of the human race. The species which we call Homo sapiens probably evolved in Africa some hundred thousand years ago. From what we know today, the way humans lived in those times did not have a stark difference from the way animals lived, until the "modern humans" who rapidly spread from Africa into the ice-free zones of Europe and Asia, developed a language some 70,000 years ago. The development of a language was a game changer. It provided an efficient tool of communication between individuals, thereby making them into outstanding creatures. As the language developed, humans spilled into other places, and effectively conquered the world. This period of time is referred to as the *cognitive revolution*. Prior to this, humans, just like animals, gathered and hunted for food.

Retracing our steps into the annals of time, it is believed that about 10,000 years ago the *agricultural revolution* emerged. This revolution probably took place in the region which we now know as the Middle East. This was, yet again, another major transformation in the way humans lived.

Unlike in the past when humans gathered and hunted for food, they learned to *domesticate* both plants and animals, and by doing so they placed their food sources within reach.

Anthropologists theorized that as humans became smarter and skillful, they learned more and more, and came up with even better ways of domesticating plants and animals, the result of which was better, easier, and more secure lives.

Harari, in his delightful book *Sapiens: A Brief History of Humankind* [Harari (2014)], challenges this prevailing theory. He claims that the agricultural revolution did not improve the lives of humans. On the contrary, he posits that humans enjoyed a stress free and easy life before the agricultural revolution, had a larger variety of food, and were less susceptible to hunger and diseases. In fact, he claims that plants actually domesticated humans, and not the other way around. The proof: Humans are living in houses (domiciles), and wheat is not!

According to Harari, there were three major developments between 5,000 and 2,000 years ago which shaped the unification of humankind. The first was the invention of currency and trade, the second was the creation of empires, and the third was the emergence of religions.

The evolution of humankind in the last 2,000 years has been fast and furious. Much later, about 500 years ago, the *scientific revolution* came to fruition. Humans acquired an enormous amount of unprecedented knowledge. They learned to harness energy in order to gain control of the world. They developed means of transportation for land, sea, and air. They also developed means for mass production.

The past 100 years have seen the rapid, steady, and explosive growth of knowledge. The possession of knowledge motivated and drove humans to go into studies of the microscopic world of atoms and molecules on one hand, and the vast regions of the macroscopic space on the other. Today, we understand the world we live in far more, and far better than our immediate ancestors understood, or even dreamed of ever understanding. The question of whether the quality of life of today's humankind has improved as a result of these revolutions is a subject of vigorous debates.

Life expectancy has increased and is still steadily increasing. Many diseases and plagues were eradicated, and as a result of technological breakthroughs products such as telephones, televisions, and computers are staples nowadays. With the advent of mobile phones, social networking, laptops, and tablets, the world is virtually within reach and is just a click away.

But there are always two sides to a story, or two sides to a coin. With all the positive advantages and developments comes a baggage of negative effects. It seems that these developments always come as a package deal — the good and the bad. Sometimes the negatives far outweigh the positives, as in the case of weapons of mass destruction, which have the capacity to blow everything into smithereens. The air we breathe and the water we drink are getting increasingly polluted and toxic. Moreover, we are still in the dark about the long-range effect of the new and genetically engineered food we eat. We are also enamored of our gadgets without really knowing how much damage they can cause.

Where all these developments will lead us, only Time will tell....

Before moving on to the next history, the reader should pause to ponder: What are the "ingredients" of the history of humankind, or of anything else? As anyone who has studied History can attest, the history of humankind could be presented as a list of events occurring at different dates and locations (the French Revolution started in 1789 and ended in 1799, in France; World War II started in 1939, in Europe, and ended in 1945, almost throughout the world). This manner of telling history is tedious. On the other hand, one can "embellish" the history of mankind with some anecdotes, some insights into the reasons for, and the consequences of the events that have occurred, and perhaps draw some feasible conclusions for the future. In the brief history presented above, I have listed very few, key events in humankinds history, besides some comments and insights on these events.

I applaud Harari for doing a great job in this aspect. If you read his book, you will find more fascinating and imaginative *ideas* about the history of time than a mere list of facts about events occurring in time and space. When we discuss the history of an abstract concept, we will see that the major part of the history is not a list of events occurring in points of time and space, but rather about the history of ideas, attitudes and interpretations related to this abstract concept.

## 2.2 A Brief History of Arieh Ben-Naim

I am lifting this brief history of myself from my CV. I was born in Jerusalem on the 11th of July, 1934, long before the creation of the State of Israel. Figure 2.1 shows baby Arieh: aged three months, Jerusalem.

I had a colorful life as a pupil, moving from one school to another. I also spent countless days not within the confines of a classroom, but out in the streets. Every once in a while, my elder brother and I visited schools. I say "visited", because it took only a few hours before we were thrown out. I will keep the reasons close to my chest.

Similarly, my mother, running out of options but with an undeterred drive to provide us with education, did the unthinkable. She enrolled us in an exclusive school for girls. You don't need to rub your eyes — you read it right! We

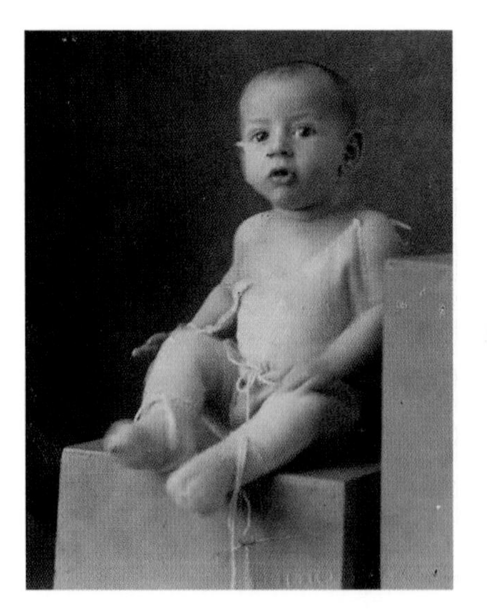

**Fig. 2.1.** Baby Arieh, aged three months.

**Fig. 2.2.** At my bar mitzvah.

"visited" a girls' school, and attended classes all dressed up in girls' uniforms. I do not remember how long it took before the school authorities "discovered" who we really were, but we were expelled, yet again from what was becoming a long string of schools.

I had my bar mitzvah (Figure 2.2) just before the Israeli independence war.

In 1949, I entered high school with virtually no solid background education. The first two years were quite tough for me, but more so for my teachers. Only in the third year did I decide to take matters into my own hands. I decided to work very diligently, burning the midnight oil in order to catch up on all those wasted years that I had spent out of the classroom. All my painstaking efforts paid off, and I graduated from high school with flying colors. In particular, I excelled in Mathematics, which became my favorite subject throughout my life.

Immediately after I graduated from high school, I entered the army, in compliance with the State of Israel's mandatory army service for high school graduates. I set my sights on becoming a pilot, and decided to join the Israeli Air Force. In those days we had British instructors, and to this day I still remember Mr. Hamilton and Mr. Bond, who were my instructors in the primary and the advanced flying school, respectively. As I look back, the years 1953–1955 were probably the most intense and happiest of my life. On May 5, 1955, at 5:55 p.m., and on the fifth day of the Jewish weekly calendar (which I, and my flying school batch mates, fondly remember and refer to as 5.5.55), I received my flying wings from the legendary, admired, and yet widely controversial Moshe Dayan, who was the Chief of Staff of the Israeli Defense Force during that time. It was a remarkable day that I will never forget (Figure 2.3).

Soon after I finished flying school, I started in the operational training unit (OTU), and flew the *Spitfire*, a British single-seat fighter aircraft that was used by the Royal Air Force during the Second World War (Figure 2.4). During that period, I had one of the most horrifying experiences of my life. While I was approaching the runway for landing, one of the pilots seemingly came out of nowhere, and entered the runway to take off. I heard the screeching voice from the control tower, commanding me with the words "Go around,

Date: 5.5.55 at 5.55 PM

**Fig. 2.3.** Receiving the flying wings from Moshe Dayan.

**Fig. 2.4.** The Spitfire.

go around, go around!". I tried to push the throttle, but the engine failed dismally. I had only a few seconds before hitting the ground, and I was sure I was at death's door. Part of me was following the instructions for an emergency landing which I had practiced so many times, and the other part of me was riveted on the thought of catching the moment of death. The wings fell apart as a result of the violent impact on the ground, creating a loud thud that I will probably never forget for the rest of my life; and a plume of dust, resembling a huge cloud, and partially obscuring my vision, fell on me like a shower. Then came a deafening silence, and numbness, and I was not certain whether I had survived or not, or whether I had lost a limb. Slowly, reality dawned — I was alive! Ironically, I made a successful belly landing in a cornfield at a nearby kibbutz. While pondering what to do next, I saw Ezer Weizman — then Chief Commander of the base, and who later became the seventh President of the State of Israel — frantically running toward me, almost resigned to the fact that he would be confronted with a lifeless body. Lo and behold, right before his very eyes was the very much alive me. He was shocked to see me alive and composed.

This incident sealed my fate — I decided flying was not for me, and stopped training in the OTU. I was sent to a navigator's school, flew old, two-engine planes, and after about a year, I was released from the army.

I started to learn chemistry in the Hebrew University in 1957, not because I liked it but because my chemistry teacher, Avner Treinin, convinced me to take that subject instead of mathematics (which was my first love).

I was bored stiff during the first two years of chemistry courses and laboratory experiments. My salvation from the pit of boredom came when I decided to focus on physical chemistry and afterward on statistical thermodynamics.

I was a freshman in chemistry when I got married.

I did my master's degree on the thermodynamics of solution of noble gases in water, which was largely experimental. I tried developing a method for measuring the solubility of argon in water — but I failed. I continued this endeavor into my Ph.D. work under the tutelage of Shalom Baer.

My relentless efforts paid off when, in 1964, I solved the problem. I developed a very accurate method for measuring the solubility of argon in water (Figure 2.5). I have been proud of that achievement until today.

Toward the end of my Ph.D. work, I focused on two problems: the structure of water, and the entropy of solution of argon in water. These two subjects were very difficult — and I spent much of my academic years in research aimed

**Fig. 2.5.** The apparatus I built to measure the solubility of gases in water.

**Fig. 2.6.** My Ph.D. lecture in 1964.

at understanding the structure of water and the unusually large and negative entropy of solution of noble gases in water. My Ph.D. lecture was given in 1964 (Figure 2.6).

While doing my Ph.D., I also took courses in mathematics: algebra, geometry, probability, functional analysis, and many more. In this way, I was able to go back to my favorite subjects, which in turn helped me in my research work on theoretical problems in the theory of liquids and aqueous solutions.

I started my postdoctoral work on the theory of ionic solutions at Stony Brook University in 1965. I did not particularly like the subject and barely after a year, I left the university and moved to Bell Telephone Laboratories at Murray Hill, New Jersey. It was there that I worked on the theory of liquid water, a subject that occupied most of my research years.

In 1969, I returned to the Hebrew University as a senior lecturer. It was at one of the Gordon conferences on water that the then president of Plenum Press approached me about writing a book on the theory of water. I experienced a myriad of emotions — shocked, daunted, flattered. Surprised at his offer, I asked him why he had singled me out — an inexperienced researcher who had just finished his postdoctoral studies. His answer boosted my confidence — he had sought the counsel of several people, who recommended me to do the job.

It took me about three years to write the book, juggling my time between my teaching duties, writing, and doing research work.

My first book was published in 1974. Two years later, I learned about the problem of hydrophobic interactions. These were some mysterious, strong interactions between simple solutes in water. Since I was an "expert" in the theory of aqueous solutions of noble gases, I was asked to write a review article on it. I met Walter Kauzmann on several occasions; he was instrumental in encouraging me to write my book on water, and shared with me his ideas about the so-called hydrophobic effect.

Basically, the hydrophobic effect was postulated to explain the stability of proteins. There are, in fact, at least two different hydrophobic effects: solvation and interaction (Figure 2.7). Kauzmann's idea was that an inert group on a protein will tend to avoid water and hence enter the interior of the protein. Similarly, two simple nonpolar solutes interact

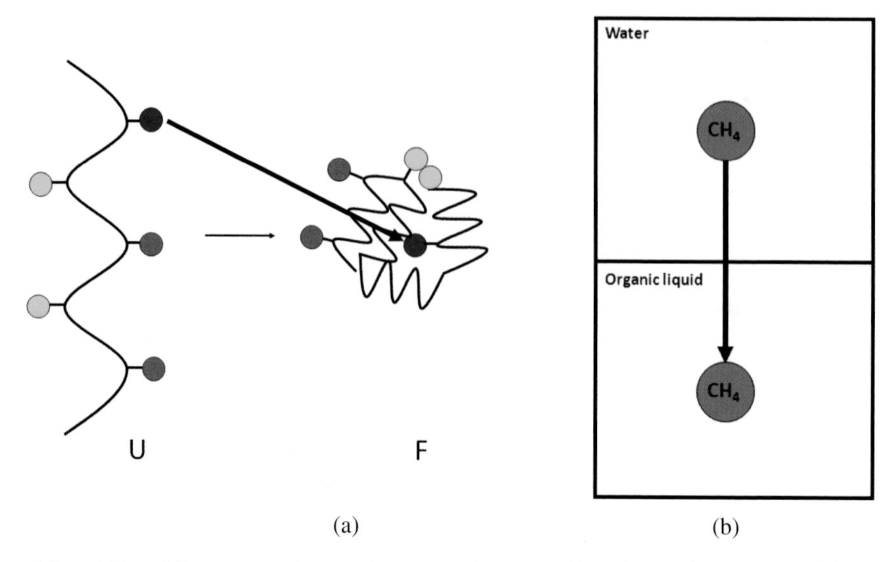


**Fig. 2.7.** Kauzmann's conjecture; the transfer of a polar group (blue circle) from being exposed to water into the interior of a protein (a) is modeled by the transfer of a nonpolar solute from water into an organic liquid (b).

weakly when they are in the gaseous phase, or in any liquid, whereas when the same solutes are in water the intermolecular interactions become stronger. I was fascinated by this problem (Figure 2.8), and I spent almost 15 years studying it, and in 1980 I wrote the book entitled *Hydrophobic Interactions*, published by Plenum Press.

The main mystery of the hydrophobic effect is this:

Why do two simple, inert, and innocent atoms like argon attract each other farmore strongly in water than in any other known liquid?

Kauzmann postulated that these interactions could explain the stability of protein. Later, people extended this idea to explain the fast folding of protein. When protein is

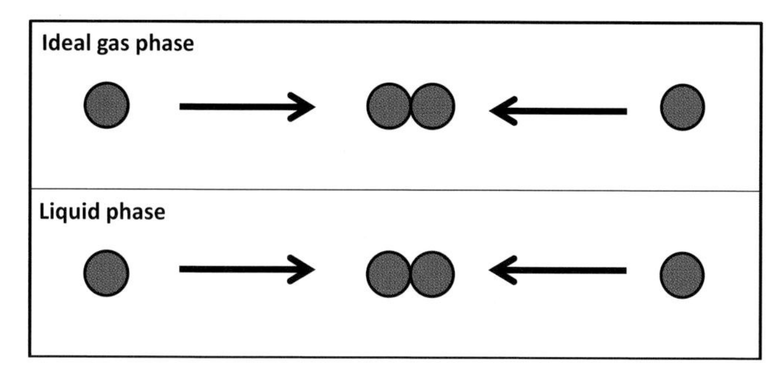


**Fig. 2.8.** Hydrophobic interactions. Two nonpolar molecules are brought from infinite separation to a close distance, once in an ideal gas phase and once in the liquid phase (water). The difference in the strength of the interaction is a measure of the hydrophobic interaction.

synthesized, it is a sequence of amino acids, much like a linear sequence of beads. How does the protein "know" how to fold into a very specific 3D structure, and fold very quickly? Most people (including me) believed that the hydrophobic effect was responsible for this process — and concluded that this effect is the *most important* one in biological systems, and to life phenomena in general.

It is surprising that many scientists spent their whole career studying, teaching, and researching the hydrophobic effects without questioning the basic tenet of the dominance of these effects in biology. In the late 1980s, while spending a year at NIH, Rockville, Maryland, I examined the whole question of the role of water in biological processes. To my surprise, I discovered that there are other effects, which I referred to as hydrophilic effects, and which might be more important in biological systems.

This revolutionary idea was not easily accepted. Many scientists' minds were fixated on the idea of hydrophobic effects, and either refused to accept or were unable to understand the hydrophilic effect. The struggle to convince the biochemical community of the importance of the hydrophilic effect is still far from over.

During my two-year stay in Burgos, Spain, I read many popular science books on entropy and the second law. Most writers claimed that the concept of entropy is the most mysterious concept in science. I did not share that belief. In 2007, I published my book *Entropy Demystified*, where I try to dispel the mystery surrounding the entropy. The history of this book is narrated in Section 2.3. Since then, I have published three other books on entropy and the second law.

I also continued my research on water and aqueous solutions, and published two books in 2009 and 2011 on these two subjects.

Since retiring in 2003, I have spent much of my time reading popular-science books on various topics in science. Among the many books I have read is *A Brief History of Time*, written by Hawking in 1988. I found the book quite boring, unclear, and unconvincing. In 2005, a new version of the book was published (with Mlodinow) — *A Briefer History of Time*. Although it is an improved version of the original book, I did not like it.

In early 2015, I decided to write a new book, to be entitled *The Briefest History of Time*. That is what I am writing now.

Before you continue, take a pause, go through "A Brief History of Arieh Ben-Naim," and make some notes as to how much of this section is a genuine *history*, and how much of it is not.

## 2.3 A Brief History of the Book *Entropy Demystified* [Ben-Naim (2007)]

Here is another history of something written for the first time in history. As in the case of Section 2.2, I have the privilege of knowing the history of this book better than anyone else....

I conceived the idea of writing the book *Entropy Demystified* while I was in Burgos, Spain. I read many popular science books, some of them touching on the concept of entropy and the second law. I did not like what I read. In particular, I did not like authors' referring to entropy as being the most mysterious concept in physics. For instance, Greene (2004) writes:

> And among the features of common experience that have resisted complete explanation is one that taps into the deepest unresolved mysteries in modern physics, the mystery that the great British physicist Sir Arthur Eddington called the arrow of time.

I felt that I could explain the concept of entropy in simple language and dispel the mystery surrounding it. That was how the title "Entropy Demystified" was conceived. The actual writing of several drafts took place in Burgos, Spain and ended in La Jolla, California. This phase belongs to the prehistory of the book. The actual history of *Entropy Demystified* was in July 2007. I do not know exactly when it was printed in Singapore — I was not physically present during that stage. I received the first copy of the book in July 2007 (Figure 2.9). A few years later, it was translated into Italian, Spanish, Japanese, and Chinese. I liked very much the cover

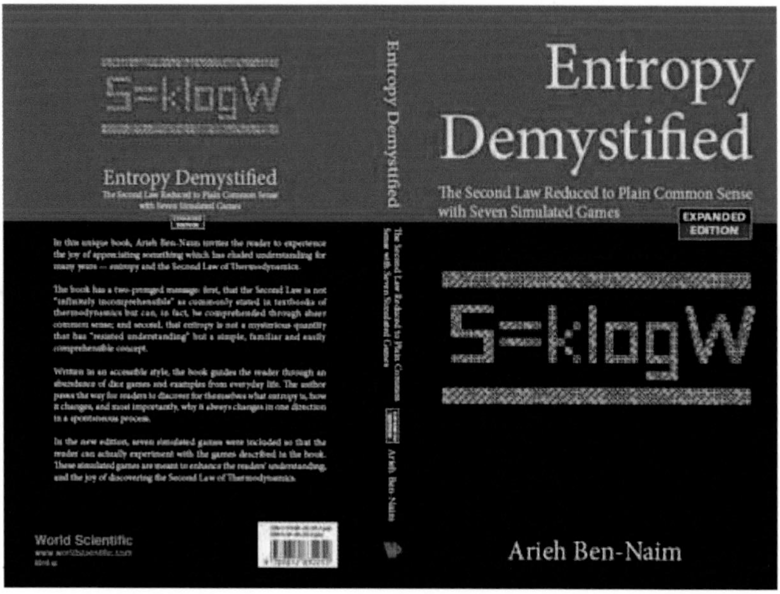

**Fig. 2.9.** The cover of the book *Entropy Demystified.*

design. I also enjoyed (and still do) the smooth and nice feel of the cover as I ran my fingers along it. Most important of all, sometimes when I read parts of the book, I cannot believe that I wrote it. For someone like me who is not a native English speaker, and whose background in school as a young boy was very poor, I feel that I deserve a pat on the back. Actually, the credit for the English is due to my wife, Ruby.

I sent copies to a few friends and colleagues, who read the book and sent me valuable comments.

In October 2007, Diego Casadei sent me a copy of the review he had written on the book. I was ecstatic about and proud of the review.

Right after its publication, my son, Yuval, prepared seven simulated games to help the reader understand the "experiments" described in the book. In 2008, an expanded edition of the book was published. A revised edition was published in 2012, wherein a few errors were corrected.

Through the years, the sales of the book grew steadily. It did not sell millions — over 4,000 copies, in fact. What was more gratifying was the few hundred emails sent by scientists and nonscientists alike, who expressed how much they had enjoyed reading my book. Some emails were as simple as "Thank you for writing the book," but were nevertheless heartwarming. Getting very positive feedback from readers boosted my enthusiasm to write more books — a luxury I did not enjoy before I retired.

Most of the reviews of the book were favorable. Some were quite brief, some lengthy. I even made friends with a number of the readers, some of whom I have met. I always enjoyed reading the reviews.

But, as the saying goes, you can't please them all. A nasty review was posted on Amazon.com, which prompted me to respond both on Amazon and in one of my more recent books, *Entropy and the Second Law*, published in 2011.

At one of the conferences that I attended in 2008, someone told me that this book can be downloaded for free from the Internet. My initial reaction was that the practice would affect the sales of the book. In retrospect, however, I also gained some measure of satisfaction upon hearing the "news." First, because someone found the book sufficiently interesting to take the trouble to upload it on the Internet; second, I believe that most people who download the book for free may not have the funds to buy a copy. Therefore, I should be happy that those people have the chance to read, and hopefully enjoy reading, the book.

Today, almost eight years to the day when it was first published, I still get emails from people all over the world. It always warms my heart to read their emails, and this for me is far more important and gratifying than earning from book sales. Indeed, sometimes there are things that money cannot buy. I am glad that I wrote the book and I hope it has contributed to demystifying the concept of entropy and the second law.

Before moving on to the next history, I would like to suggest that the reader go through this section and make some notes as to how much of what I have written is relevant to the *history* of the book, and how much is not relevant to the history of the book *Entropy Demystified.*

## 2.4 A Brief History of My Writing Instrument

The pen I am holding right now (Figure 2.10), a conduit that transports the thoughts and ideas from my brain, down to my hand, and finally onto paper, was probably "born" sometime

**Fig. 2.10.** The pen. The history of it is told in Section 2.4.

in a factory in China during the last few years. I am clueless as to how it was produced, and when exactly it was transformed from the raw materials and then injected with ink to make it functional. It was probably exported to Israel in 2014, after spending a few months at sea, and being lulled by the waves as it lay in the berth. It found its way to the ports of Israel, and then into the Academon, a store on the Hebrew University campus, and one day I walked into the store, picked it up, and decided to buy it. That was in April 2015. I put it to good use as soon as I bought it.

I like the way the rolling ball runs smoothly, almost caressing the paper as I write, as my thoughts are transported from my brain, down to my hand, and to this humble pen, which translates my thoughts into text. I also like it that the transparent casing allows me to see how much ink is left. As I continue to write every day, with the same bravado and enthusiasm as for all my books, the ink will be used up in just a couple of days. I will throw the pen away, and someone will sort out the garbage in the dump site, separating recyclables from non-recyclables. Being made of plastic, it will probably end up being recycled. Then, its "history" will come to an end.

I am no soothsayer, and so I cannot foretell when the pen will breathe its last. I am so attached to it by now that I do not even want to imagine how it might meet its grim end. I hope its history will come to an end before it piteously crackles as its body is eaten up by flames in the garbage incinerator.

Now, please read the last sentence of Section 2.3, and also do what I suggested there for this section.

## 2.5 A Brief History of Beauty

In the previous sections, we narrated the histories of some substantial objects. All these histories can be presented as a series of *events* occurring at some points in *space*, and at some *points* in time.

When we contemplate narrating the history of an abstract concept, we face a serious difficulty. Can we narrate the history of beauty by listing a sequence of events which beauty has gone through? Was beauty created (or "born," or made to appear) in some location, and at some point in time?

What does one usually mean when one says that a history of beauty, or any other abstract concept, is a history of ideas, views, attitudes, etc. about or toward that concept? There is no doubt that different cultures have different views about beauty, and these attitudes have evolved with time at different places. Even this type of history does not have a clear-cut beginning, and certainly it is unclear whether it will have an end. In ancient times, kings and queens were buried with their ornaments, and sometimes even some unfortunate slaves were buried with them, as in the case of Egyptian pharaohs who were accompanied by slaves. The belief was that the slaves would serve their masters in the afterlife. These kings and queens also wore amulets, which they believed would enhance their power upon resurrection, and in their next life. In Greek mythology, Aphrodite is the goddess of love, beauty, and procreation. In Roman mythology, Venus is the goddess of love, sex, beauty, and fertility.

In many instances, the Bible describes people as beautiful — Rachel is described as "shapely and beautiful," her son, Joseph, as "well-built and handsome," and King David as "ruddy, bright-eyed, and handsome." It is interesting to note that in the "Book of Proverbs" beauty is considered vanity, and for a woman of valor, wisdom and good deeds are prerequisites, not beauty. Of course, it is well known that the notion of beauty varies from culture to culture, and in the same culture from one period of time to another. All these interesting facts cannot be considered to be parts of the history of beauty. Equivalently, we can say that beauty does not have a history and, in short, beauty is a timeless concept.

## 2.6 A Brief History of a Mathematical Theorem

To be specific, consider the Pythagorean theorem in geometry. This theorem states that in a right-angled triangle the area of the square on the hypotenuse is equal to the sum of the areas of the squares of the two sides of the triangle (a and b in Figure 2.11). (In this section, I am referring to the theorem in Euclidean geometry only.) Was the theorem "born" when it was first described, or proven by Pythagoras? Or perhaps it was known before Pythagoras but was never formulated or proven by anyone? Neither proving nor announcing nor publishing the theorem can be considered to be the history of the Pythagorean theorem. Here, we touch upon this long-debated question: Is a mathematical theorem "invented" or just "discovered"?

Certainly, we feel that the Pythagorean theorem *existed* long before it was discussed by Pythagoras. In fact, it is inconceivable to claim that the theorem came into existence

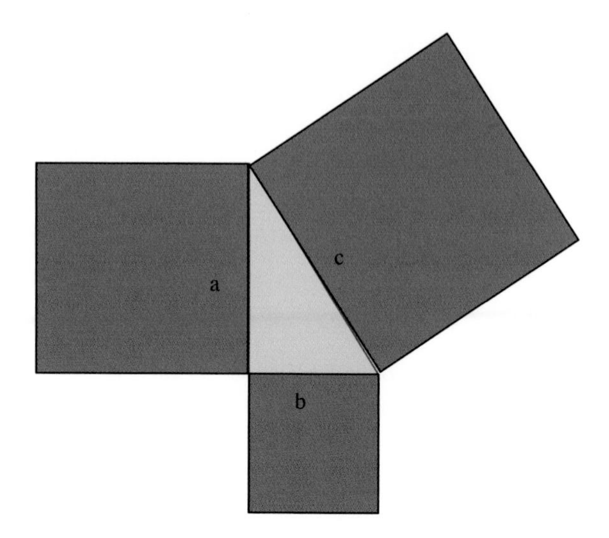


**Fig. 2.11a.** Pythagoras' theorem.

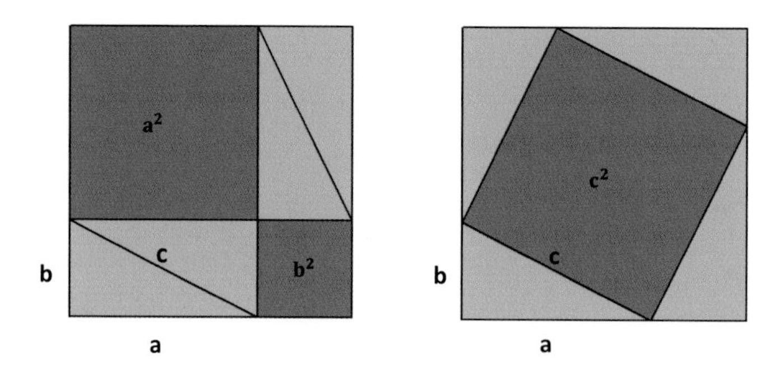


**Fig. 2.11b.** The visual proof.

(was born, created, or made to appear) at any specific point in time, or any specific location. It is also inconceivable to claim that the theorem will ever die (or disappear or be destroyed). The validity of the Pythagorean theorem has no boundaries in time, or space; it even does not depend on the existence of humans. In short, we can say that the theorem is timeless (i.e. it has no history). This is true of any geometrical theorem, the truth of which is independent of the existence of humans. The same must be true of the theorem that there are an infinite number of prime numbers. (Can one claim that once upon a time there were a finite number of prime numbers, and then a new theorem was "born," stating that there were an infinite number of prime numbers?) Similarly, the theorem that $\sqrt{2}$ is an irrational number cannot have a history. (We cannot claim that once upon a time $\sqrt{2}$ was rational number, and then one day it became an irrational number….) Perhaps some mathematical theorems in some abstract spaces (Hilbert or Banach spaces, etc.) which were *invented* by mathematicians can be claimed to begin at some point in time by some mathematician at some locations in space.

Note that a Hilbert space is not a "space" — it is a collection of all functions having some properties. Mathematicians like to use the word "space" for a collection of objects between which some rules are defined. A Hilbert space is a *creation* or *invention* of humans. It is difficult to claim that a theorem proven in a Hilbert space has *always* existed, as one could easily do for the Pythagorean theorem in geometry. In other words, if humans never existed and Hilbert space was never defined, could one claim that a theorem in Hilbert space always existed?

Similarly, there is a whole branch of mathematics called complex analysis (or the theory of functions of complex variables) which deals with complex numbers involving the imaginary number $i = \sqrt{-1}$. This number is not real; there is no real number $x$, the square of which $(x^2)$ is a negative number, say $-1$. Again, any theorem in complex analysis can be said to have a beginning when it was first proven. If the theorem is correct, it would be difficult to claim that such a theorem will ever come to an end (or die). A most remarkable identity involving the imaginary number $i = \sqrt{-1}$ is the Euler identity, which is a special case of the Euler formula[2]:

$$e^{i\pi} + 1 = 0.$$

This is a very remarkable equation, connecting the five constants $e$, $\pi$, $i$, 0, and 1; $i$ is an imaginary number, $e$ the base of the natural logarithm, and $\pi$ the ratio of the circle's circumference to its diameter, $\pi \approx 3.14159\ldots$). Although this formula is attributed to Euler, it was probably known before he actually wrote it.

For our purpose in this section, can we ask about the history of this identity? Did the identity exist before any human being wrote it down? In particular, one can ask whether it is meaningful to ask about the *history* of the number $i$, which is not a real number. Clearly, the written formula had a beginning, but will it ever come to an end? Even if the universe comes to an end, I claim that this beautiful identity will survive! Thus, the whole history of this identity consists of at most, one event; its birth. Even this single event is subject to personal views.

## 2.7 A Brief History of a Law of Physics

The case of a history of a law of physics is different from the case of a mathematical theorem. For concreteness, suppose we focus on Newton's law of gravitation. This law states that the force of attraction between any two bodies is proportional to the mass of each particle and inversely proportional to the square of the distance between the two bodies. The proportionality constant is known as the gravitational constant, denoted $G$.

Unlike the case of a mathematical theorem which we discussed in Section 2.6, here we can talk about the history of Newton's law of gravitation, as well as the history of our understanding, views, attitudes, etc. regarding the theory of gravitation.

One can claim that Newton's law of gravitation, as formulated by Newton, *started* at some point in time, say during the Big Bang, at the same time as all the laws of physics were "created" or started to be operative. One can also talk about the history of the law itself. For instance, the gravitational constant might not be an absolute constant. It might have changed with time. It could be that these changes in the gravitational constant are so small, say a billionth of one percent in a billion years, that we could not detect it. But if such change did occur, then it is appropriate to talk about the *history* of Newton's law of gravitation. It might also be the case that not only does the gravitational constant change with time, but also the way the force depends on the mass and the distance might change with time. Such changes of the law could also be reckoned with as part of the history of the law of gravity.

Finally, it is in principle possible that at some point in time the universe will *crunch* into a single point, referred to as the Big Crunch, and all the laws of physics, including Newton's law of gravitation, will cease to be operative. We can say that this would *end* the history of Newton's law of gravitation.

Thus, although we do not know anything about the history of Newton's law of gravitation, such a history could, in principle, exist.

It should be emphasized that we talked about Newton's law of gravity, and not about the generalization of the theory of gravity developed by Einstein. This generalization belongs to the history of the theories of gravity, rather than the history of Newton's law of gravity.

In modern cosmology, people have speculated about the beginning of the universe at the Big Bang, and the end of the universe at the Big Crunch. Even if such events have any reality, it is not clear that the birth (and the death) of the universe is the same as the birth (and the death) of space and time (see Chapters 3 and 4). Did beauty *begin* at the Big Bang? Was a mathematical theorem, or a law of physics, born at the Big Bang?

These questions remind me of a story I heard a long time ago in a calculus course.

A famous French mathematician proved an extraordinary theorem. The theorem states that if a function has some properties, *a*, *b*, and *c*, then it must also have the property *d*. The proof of the theorem was not a trivial matter, and many mathematicians hailed the theorem as the most elegant, profound, and potentially useful theorem. No one ever found any error in the proof, and its validity was beyond question.

One day, a young student who wanted to use this theorem to prove another theorem in another branch of mathematics found, to his astonishment, that there is *no function* which has the properties *a*, *b*, and *c*, as were assumed by the French mathematicians who published the theorem.

The student concluded that the theorem is indeed *exact.* However, if there exists no function for which the theorem can be applied, then one can say that this theorem is empty or void.

Cosmologists tell us that the laws of physics *started* at the Big Bang, and that before the Big Bang *nothing* existed. If nothing was "there," on which a law of physics operated, was the law empty or void?

## 2.8 A Brief History of the Theory of Evolution

Before we tell the history of the *theory of evolution*, we must clarify that we are referring here to the Darwinian theory of evolution, not to any other mechanisms of evolution, and not to the history of the different theories about evolution, and not about *evolution* itself. In fact, we treat the theory of evolution like any law of nature.

Evolution (in the Darwinian sense) probably *began* whenever molecules started to reproduce and when the random mutations of these molecules occurred. Thus, the birth of evolution is not necessarily identical with the birth of biology, or the birth of life on our planet, or in any other part of the universe. It is difficult to pinpoint the time (as well as the location) at which the chemical evolution turned into biological evolution. This aspect of evolution is discussed in great detail in Pross' book (2012) and commented upon by Ben-Naim (2015). Note carefully that so far we have talked about *evolution*, not the *theory of evolution*, which deals with the *mechanism* of evolution. The *theory* of *evolution* itself, whether you accept it or not, does not change with time; it is timeless.

Of course, during *evolution* new molecules were created and new species evolved, but the *theory of evolution* itself did not change. The second important event to be reckoned with as the *history of evolution* might be its demise. Will it ever cease to operate? Perhaps yes perhaps no. Perhaps the whole universe will reach a state of thermal equilibrium, then it will be meaningless to talk about *evolution* of life in the universe. Whatever will be the *fate* of *evolution*, the *theory* of *evolution* will never come to an end; it is timeless. This is the same conclusion we reached in connection with a mathematical theorem or a law of physics. *Evolution* cannot operate in a world having no chemistry or biology, but the *theory of evolution* is independent of the existence of molecules or animals.

# 3

# The Briefest History of Space

We are all familiar with the three-dimensional (3D) space we live in. There is no need to define this 3D space in terms of other, more fundamental concepts. Space is an abstract concept; we cannot see it, touch it, or feel it with any of our senses. On the other hand, space is a very real place where all objects we see and touch are located. What is the history of the 3D space? Some physicists will tell you that the *space* was "born" or started during the Big Bang. During that time, Time was also created. But *where* was space born? If we are going to construct the history of space we have to list a sequence of events through which space passed at some points in Time, and at some points in space. But we are talking about the history of the very concept of space, not about some specific points in space, and not about ideas about space. If we are to register the location of space's birth, then we must assume that there is a Super-Space, denoted S-space, on which we record the history of space. Figure 3.1 shows schematically the super space on which we register a few events which are reckoned as the history of space.

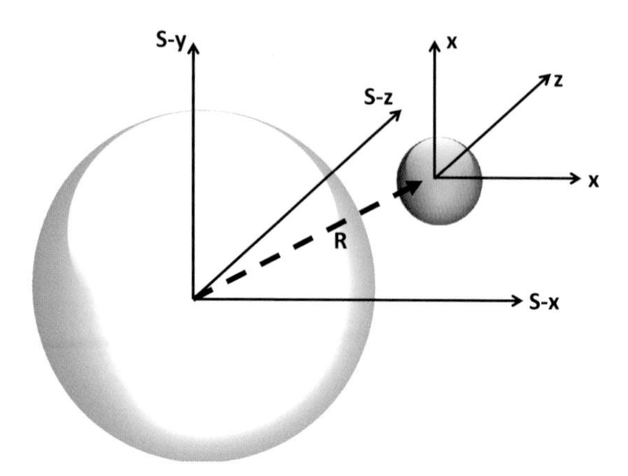


**Fig. 3.1.** The space and the S-space.

Are there any other events that belong to the history of space? Perhaps, its expansion could be reckoned as part of its history. Perhaps its eventual death, as is speculated at the Big Crunch. What about any other events? Any suggestions? At the moment we cannot say much about the history of space. Note carefully that if space *was once absolute*, and at some point in time it *became relative*, then we could consider these events as genuine parts of the history of space. This is not the case, however. Space did not change from being *absolute* to *relative*; it is we, human beings, who changed our view about space.

Some physicists will tell you that it is meaningless to ask *where* space was created or born, because space did not exist before the Big Bang. In my opinion, it is meaningless to claim that it is meaningless to ask where space was born. I cannot even imagine what it means when one says that *before* the Big Bang space did not *exist*. Regarding the question of the meaning of "the time *before* the Big Bang," see Chapters 4 and 6.

# 4

# The Briefest History of Time

Once upon a time, Time was born. It was born at *no-time*, since at that time, time did not exist yet. Of course, we cannot tell the exact date of its birth. Physicists believe it happened approximately 13.7 billion years ago, but note that days, months, and years did not exist at that time. Right after that "big event," nothing had happened to Time. Time did not do anything to anything (it did not go to school, nor was it thrown out of school, it did not get married, never traveled in space, did not go through hard times or bad times....). Also, no one did anything to Time. To the best of my knowledge (and I presume also to the knowledge of all scientists), no other events associated with Time happened at some points in Time, at some points in space. Note that if Time *was once absolute*, and at some point in time it *became relative*, then we could consider these events as genuine parts of the history of Time. This is not the case, however. Time did not change from being *absolute* to *relative*; it is we, human beings, who changed our view about Time.[3]

Before we end the history of Time, we should mention that perhaps one day, just after sunset, Time will "fade," or come to an end. It is impossible to tell the exact date of this event, or whether such an event will ever occur. Therefore, the only event which may be reckoned as the history of time is its birth. And even this event is highly speculative.

Here, we ended the history of Time. It is the *briefest*, compared with *Brief* and *Briefer*, but it is also the *longest* history of Time. Perhaps, Time does not have any history — if this is the case, then Time is *timeless*!

Most cosmologists who discuss the history of Time are actually discussing the history of ideas about time, theories of time, and so on. We will discuss some of these ideas in the following chapters of this book.

As we have noted in Chapter 3, people connect the Big Bang with the beginning of Time, and claim that it is meaningless to ask what happened *before* the Big Bang.

Recall that for any history of X, we can ask what happened before X was born (or created, or began). Only when X is Time itself are we forbidden to ask what happened *before* Time began. In my view, it is meaningful to ask what happened *before* the Big Bang, as well as *after* the Big Crunch (presuming that these events have any reality). In other words, I think that it is meaningless to claim that "it is meaningless to ask about the time preceding the Big Bang."

In connection with the history of space as discussed in Chapter 3, we asked the following question: At which specific point of space was space created? Similarly, when we discuss the history of Time, it is legitimate to ask: At which point of

Time was Time born (or created, started, etc.)? Disregarding the question about the location in space where Time was born, it is clear that the time in which Time was born cannot be registered on the same Time axis, the history of which we are narrating. Perhaps the time of birth, as well as other events which Time has gone through, should be registered on a different time axis, which we may refer to as the super-time. (See Section 1.2.)

**The History of Time**

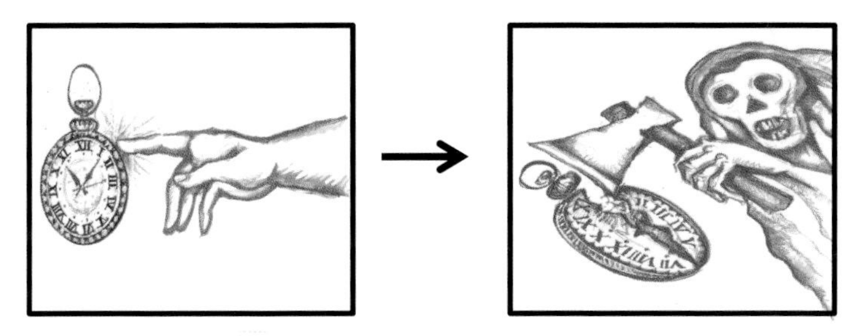

**Time's birth ----- and ----Time's death**

# 5

# Entropy and the Second Law of Thermodynamics

In this chapter, we present the concepts of entropy and the second law of thermodynamics. This chapter is not a history of entropy, or a history of the ideas about the concept of entropy. We will focus only on the bare minimum information for the layperson who wants to understand why many popular-science books connect entropy and the second law with the so-called arrow of Time. As we will soon see the idea of the arrow of Time exists only in our minds. The entropy is not a function of Time, and the second law has nothing to do with the arrow of Time. In fact, a chapter on entropy and the second law has no place in a book on Time. We bring it up here only because almost anyone who writes about Time also writes about the apparent connection between entropy and the arrow of Time (see Chapter 6).

We start with short, historical notes on entropy. We then introduce the Shannon measure of information (SMI), which is essential for understanding entropy. We will then show how entropy emerges from the SMI. This is almost a miracle. The SMI was defined in communication theory, and entropy was defined in connection with heat engines. There is nothing common between the two fields. Yet, the entropy turns up to be a special case of an SMI.

Once we are done with entropy (and I hope by then you will have a good idea what entropy means), we turn to discussing the second law. To do this, we first study some examples for which the SMI changes with time. Then, we show how the SMI of a thermodynamic system changes in a spontaneous process in an isolated system. It turns out that the value of the SMI at the *state of the equilibrium* is *maximal*, and this value is proportional to the entropy of that system. It will be clear that entropy is not a function of time. This is in stark contrast to what most people think about entropy — some even equate entropy with time's arrow.

## 5.1 Some Historical Notes

The term "entropy" was coined by Clausius in 1865. This date may be viewed as the birth date of the concept of entropy [the birth of the concept of entropy should be distinguished from what some scientists refer to as the *origin* of entropy; the latter is discussed in Ben-Naim (2015)]. Clausius, among many others, noticed that many spontaneous processes always proceed in one direction. Examples are:

1. The expansion of gas from $V$ to $2V$ upon the removal of a partition between the two chambers (Figure 5.1a);

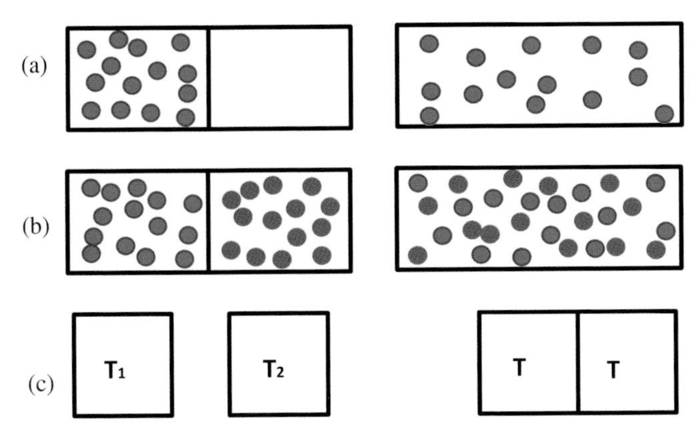


**Fig. 5.1.** Three spontaneous processes: (a) gas confined to a smaller volume will always expand and occupy a larger volume; (b) mixing of two gases after removal of a partition; (c) heat will flow from the hotter to the colder body.

2. The mixing of two gases (Figure 5.1b);
3. Heat transfer from a hot to a cold body (Figure 5.1c).

We never see any one of these processes occurring in the reverse direction. A gas filling the entire volume on the right hand side of Figure 5.1a does not spontaneously (i.e. without any intervention from the outside of the system) condense to occupy half of the region. Two mixed gases, as on the right hand side of Figure 5.1b, never unmix spontaneously into two pure gases. Heat never flows spontaneously from a body at a lower (or equal) temperature to a body at a higher temperature.

There are many other processes which we observe to occur in only one direction. The following question arises: Do all these phenomena have a common source? Is the direction of occurrence of all these processes dictated by one universal law, or perhaps each phenomenon is a result of a different law or no law at all?

Clausius introduced the concept of entropy, and the most famous quotation by him is:

> The energy of the universe is conserved.
> The entropy of the universe always increases.

A not-less-famous quotation by him is his explanation of how he chose the word "entropy":

> I prefer going to the ancient languages for the names of important scientific quantities, so that they mean the same thing in all living tongues. I propose, accordingly, to call S the *entropy* of a body, after the Greek word "*transformation.*" I have designedly coined the word *entropy* to be similar to *energy*, for these two quantities are so analogous in their physical significance, that an analogy of denominations seems to me helpful.

I have several reservations regarding Clausius' arguments in choosing the word "entropy." These are discussed in Ben-Naim (2008). Nevertheless, Clausius certainly deserved the credit for unifying all these phenomena under one law — the second law of thermodynamics. This law states that there is a quantity called *entropy* which is a *state function*. This means that when the *state* of the system is characterized by say, the energy $E$, the volume $V$, and the number of particles $N$, the entropy is also well-defined. When spontaneous processes occur as a result of removing a constraint (say, removing the partition between the two gases in Figure 5.1b), under the condition that the entire system is isolated, the entropy of the system will either remain constant or increase. It will never decrease. Two examples of "processes" for which no entropy change will be observed are shown in Figure 5.2.

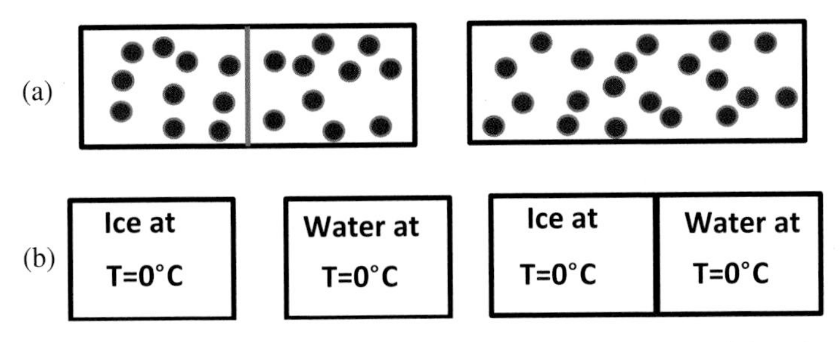


Fig. 5.2. Two processes for which no entropy change will be observed: (a) removal of a partition between two compartments separating two compartments having the same gas at the same temperature and the same density; (b) bringing into contact ice and water, at the same temperature and pressure ($T = 0°C$).

Note that the second law in this particular formulation applies to well-defined systems. "Well-defined" is with respect to the macroscopic state, not the microscopic state, of the system. There are other formulations of the second law [see Ben-Naim (2008, 2011)]. We will not need these here. It should be mentioned, however, that Clausius erred in his statement about the entropy of the *entire universe*. We will further comment on this in the following sections.

Clausius did not provide any method for calculating a numerical value of the entropy of a specific system. Instead, he defined the entropy change for one specific process. When a small quantity of heat (denoted $dQ$) is transferred into a system at a given temperature $T$ ($T$ is the absolute or the Kelvin temperature), the entropy of the system will increase by the amount $dS = dQ/T$ (Figure 5.3a). When a small quantity $( - dQ)$ is *removed* from the system being at a constant temperature $T$, the entropy of the system will *decrease* by the amount $dS = -dQ/T$ (Fig. 5.3b). (We use

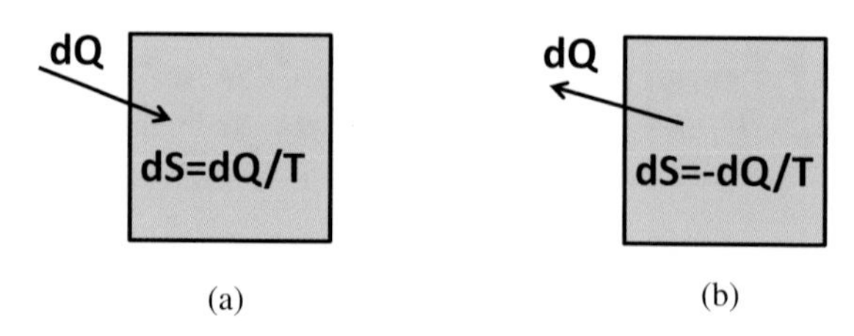


**Fig. 5.3.** (a) Adding heat $dQ$ to the system at a temperature $T$, the entropy change of the system is $dS = dQ/T$; (b) extracting the same amount of heat from the system causes a change in entropy, $dS = -dQ/T$.

$dQ$ for a very small quantity of heat being transferred to the system. We do not imply that $dQ$ is a differential of a function $Q$.)

Clearly, this formula for calculating the entropy change is valid for this particular process. Experts in thermodynamics developed an ingenious way of calculating the entropy change between any two *well-defined* states of the system. "Well-defined" means "well-defined thermodynamically." We cannot calculate or measure the entropy change when a mother gives birth, or when a person dies.

Any of the processes in Figure 5.1 may be viewed as a change from one well-defined state to another well-defined state. In the first two there is no heat transfer involved, yet one can devise a process which will carry the system from the initial to the final state, and use Clausius' formula, $dS = dQ/T$, to calculate the change in the entropy of the system.

Thus, thermodynamics is able to calculate entropy changes from one well-defined state to another. However, it is unable to determine an absolute *value* of the entropy of a system at a specific well-defined state. It is only with the help of the third law of thermodynamics that one can assign absolute entropy values to some systems (those systems which at absolute zero temperature have zero entropy).

Furthermore, thermodynamics does not provide any molecular interpretation of entropy. This fact does not diminish the importance of entropy in thermodynamics. As you may know, temperature also does not have a molecular interpretation within thermodynamics.

The first molecular interpretation of entropy was given by Boltzmann, who related the entropy of a thermodynamic system to the total number of microstates of the system. This relationship is valid for a system all the microstates of which have equal probabilities (see below for some simple examples of such a relationship).

During the years, many interpretations of entropy were suggested: entropy as a measure of disorder, entropy as a measure of spreading of energy, entropy as a measure of freedom, and many more. None of these descriptions of entropy were ever proven to be correct. In some cases, entropy changes are correlated with increase in disorder, or spreading of energy or degree of freedom, but these correlations do not apply in general. In some popular-science books, one might find "examples" of the second law, such as "A children's room gets disordered with time" or "Kitchens tend to get messier with time." Such statements are not true in general, and even when they are true they have nothing to do with the second law.

In my view, the best way to understand entropy, which also dispels all the mystery surrounding entropy, is the one based on the Shannon measure of information. In the following sections, we present the definition of the Shannon measure of information, and outline the derivation of entropy from the SMI. A more detailed derivation may be found in Ben-Naim (2008, 2011).

## 5.2 The Shannon Measure of Information

In this section, we introduce an important quantity which is the cornerstone of information theory. This is the Shannon measure of information (SMI). We will present the SMI in a qualitative manner just enough for understanding how entropy may be interpreted in terms of the SMI. We will proceed to develop this concept in the following steps.

First, we start by playing a simple and familiar 20-question (20Q) game. I choose an object out of $n$ possible objects, and you have to find out which object I chose by asking binary questions, i.e. questions which are answerable by either "Yes" or "No." We assume in this game that the probability of choosing the object out of $n$ objects is the same for each object and is equal to $1/n$. We will refer to this game as the *uniform probability game* or, simply, the uniform game. To make the game more precise and also to ease the generalization to the nonuniform game, think of the game shown in Figure 5.4a. You are shown a board which is divided into $n$ regions of equal areas. You are told that a dart was thrown by someone in a blindfold, and hit a point on the board. You have to find

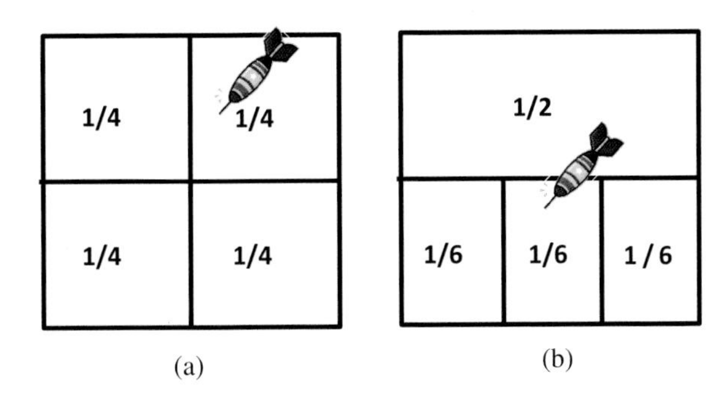


**Fig. 5.4.** Two games: (a) uniform game and (b) nonuniform game.

out in which region the dart hit the board by asking binary questions. I trust that you can easily play this game.

Second, we generalize the simple 20Q game described above to a more complicated game. The objects are chosen with different probabilities. To understand this more general game, consider a board on which a dart was thrown. The board is divided into $n$ regions of *unequal* areas, and the dart was thrown by someone in a blindfold. Now, you are told that the dart did hit a point on the board. You are also informed about the relative areas of all the regions in Figure 5.4b. Your task now is to find out in which of the regions of the board the dart hit by asking binary questions. You should realize that this game is different from the previous game because the "events" here are not equally probable. We refer to this game as the *nonuniform game.*

A quick question before we proceed. Suppose you are asked to play the 20Q game on the board of either Figure 5.4a or 5.4b. You have to pay \$1 for each question you ask. Once you find out where the dart hit, you get a prize of \$20. Which game would you choose to play?

Now, we turn to the third, and last, step. If you understand the 20Q game, and if you can answer my question above (about the preferable game out of the two in Figures 5.4a and 5.4b), you should realize that in playing the 20Q game you *lack* information on the location of the dart. By asking binary questions you *gain* information from each answer you get (and you have to pay to get that information). Eventually, you will obtain the *information* you need.

The last step does not introduce any new *generalization* to the 20Q game. It only makes the same game very large — very much larger than the games you are used to playing, or the games shown in Figures 5.4a and 5.4b. The game is large indeed, but there is nothing new in principle. You cannot possibly play this game, because you do not have enough time in your life to ask so many questions — but you can at least *imagine* playing such a game. It will perhaps be called the $10^{23}$ Q game, rather than the 20Q game, but you can be sure that if you only imagine playing this game, you will understand the concept which is still considered by many to be the most mysterious concept in physics — entropy. We will discuss this in Section 5.4.

### First step: Playing the uniform 20Q game

This is the relatively easy game. A dart hit a board which is divided into $n$ equal-area regions. You are told that the dart was thrown by someone blindfolded, and it hit one region of the board. You also know that the probability of hitting any one of the regions is $1/n$. This is why we refer to this game as the *uniform* game.

For $n = 8$, how many questions do you need in order to find out where the dart is? If you are smart enough, you can obtain the information on where the dart is with three questions. How?

You simply divide the eight regions into two groups, and ask: "Is the dart in the right group?" If the answer is "Yes," you divide the remaining four regions on the right into two groups, and ask: "Is it in the right group?" and so on. In this method, you will find out where the dart is with exactly three questions. This way of asking questions is referred to as the *smartest* way, or the smartest strategy. You can check and convince yourself that if you use any other strategy of asking questions you will need on average to ask many more questions. One can prove mathematically that by asking questions of this kind (i.e. dividing each time into two equally probable regions), one gets the maximum information for each answer. Therefore, you will need to ask minimum number of questions to get the information you need. Figure 5.5 shows the two strategies of asking questions for the case $n = 8$. Note that the amount of information (measured in bits) is independent of the strategy of asking questions. If you ask smart questions, you get the *same information* with the smallest number of questions. [For more details, see Ben-Naim (2012, 2015)].

Note that the number of questions (3) is related to the number of regions (8) by the logarithm to base 2. In this case, $3 = \log_2 8$ (here, the base of the logarithm is 2). You can check for yourself that for $n = 16$ regions you will need four questions. For $n = 32$, you will need five questions, and for $n = 2^k$ with an integer $k$, you will need to ask $k$

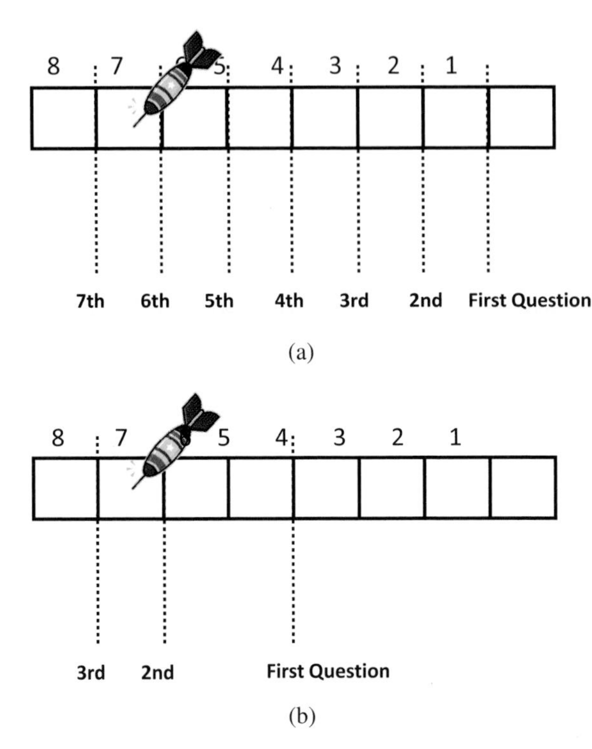


**Fig. 5.5.** Two strategies for playing the 20Q game: (a) the dumbest and (b) the smartest.

questions. Note that when we *double* the number of regions, the number of questions we need to ask (smart questions) increases by one. We can prove in general that, for any $n$, the average number of questions we need to ask is approximately $\log_2 n$, plus or minus 1. We say "average" because for any $n$, say $n = 7$, we cannot divide at each step the total number of regions into two equally probable groups. But we can try to do it as closely as possible; for instance, divide the seven regions into four and three regions. The general result for any number ($n$) of equally probable regions is that you need to ask, on the average about $\log_2 n$ questions in order to obtain the

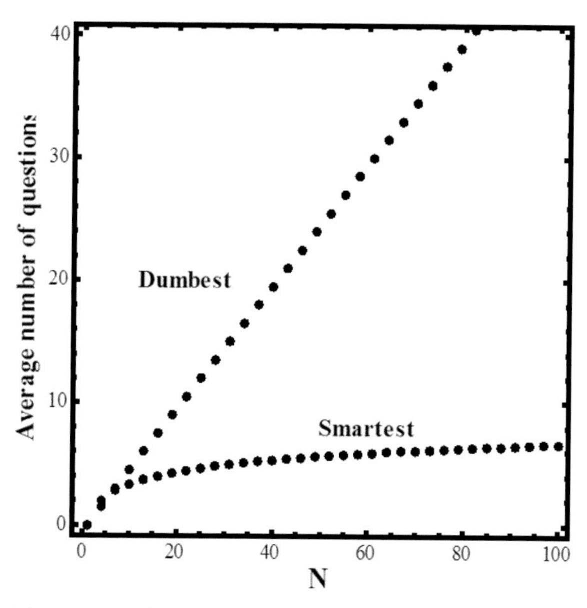


Fig. 5.6. The dependence of the average number of questions as a function of $N$ for the dumbest and the smartest strategy.

required information. All this is valid for a system of equally probable regions. It is very easy to prove that the relationship between the number of regions and the number of questions you need to ask is given by $H = \log_2 n \pm 1$. Figure 5.6 shows the average number of questions one needs to ask as a function of $n$, for the *smartest* and *dumbest* strategies. The "dumbest" strategy is when you ask systematically whether it is in region 1, in region 2, and so on. We will not need it here.

## Second step: Playing the nonuniform 20Q game

By "nonuniform game" we mean unequal probabilities to the different events. To be specific, consider the modified game shown in Figure 5.7. Here, we divide the board into eight

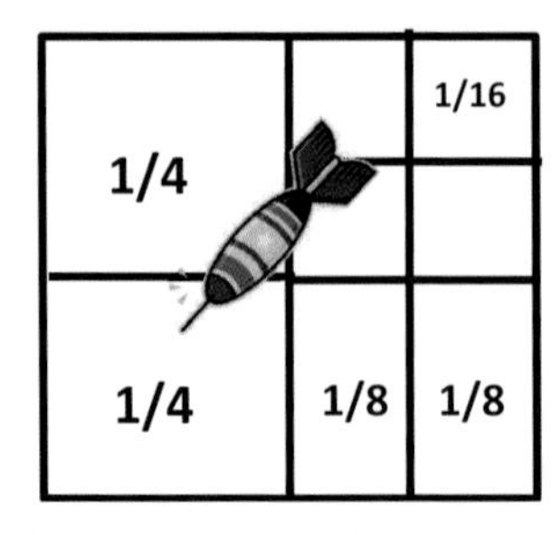


Fig. 5.7. A nonuniform game.

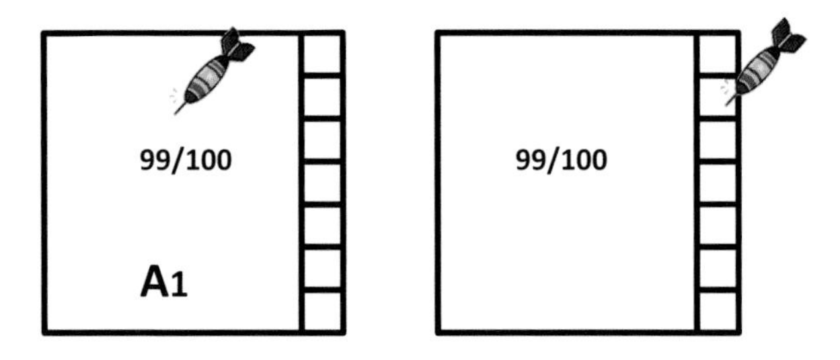


Fig. 5.8. An extreme nonuniform game.

regions. Unlike the game in Figure 5.5, here the areas (and hence the probabilities) for hitting each area are different.

Usually, when we play the parlor version of the 20Q game, we assume implicitly that the probability of choosing a specific object or a specific person is $1/n$, where $n$ is the total number of objects from which one is selected.

In the example shown in Figure 5.8, the areas of the regions are not equal. This makes the *calculation* of the number of questions more *difficult*, and yet the actual *playing* of this game is *easier* than for that shown in Figure 5.5.

One can prove mathematically that the average number of (smart) questions you need to ask in order to obtain the information (on where the dart is) is given by the famous Shannon formula $H = -\sum p_i \log_2 p_i$, where $p_i$ is the probability of the event $i$, which in our case is simply the relative area of the region $i$. We will refer to $H$ as the Shannon measure of information (SMI).

One can also prove that the quantity $H$ defined above is *always* smaller than the quantity $H$ defined on a uniform distribution with the same total number, $n$. In terms of the 20Q game, we can say that playing the game in Figure 5.7 is always *easier* than playing the game in Figure 5.5. Easier in the sense that you will require, on average, fewer questions to obtain the information you need.

This mathematical result is quite obvious intuitively. If you are offered the two games in Figures 5.4a and 5.4b, and you have to pay $1 for each answer, and get a prize x when you get the information, it is always advantageous to play the nonuniform game. You can try to play the two games and convince yourself that the uniform game always requires more questions. This answers the question we posed at the end of Section 5.2. If you are not convinced, try to take a more extreme game, shown in Figure 5.8. Here, if you know the distribution, you should ask the first question: "Is the dart in area $A_1$?" The chances of getting a "Yes" answer are 99/100; "No," 1/100. This means that if you play the game many times, you will get the information (on where the dart is) with about one question. In fact, from Shannon's formula you will find that the average number of questions is less than one. This means that even without asking any questions you will know with a high probability where the dart is. [For more details, see Ben-Naim (2008).]

## *Third step: Generalization for a 20Q game to an over $20^{23}$ Q game*

Now that you have an idea of the relationship between the number of objects and the average number of binary questions, you need to ask in order to find out which one of the objects was selected. Let us play a very large game of the same kind.

Suppose you have a single atom in a cubic box of edge d. The *state* of the atom can be described by its location and its velocity at each instance of time. Let $l$ be its location and $v$ its velocity. Clearly, the pair $(l, v)$ describes the state of this atom. We refer to this as the microscopic state, to distinguish it from the macrostate, which we use in thermodynamics. It is also clear that there are infinite states of this kind. Therefore, if I know the *exact* state of the atom, and you have to find its location by asking binary questions, you will need to ask on average an infinite number of questions.

Fortunately, there is the uncertainty principle in physics. This principle states that you cannot determine the exact location and velocity of the atom, but there is a limit to the "size of the box" $(\Delta l \Delta v)$ in which you can determine the state of the atom. This passage from the infinite number of states in the *continuous* range of locations and velocities to the finite number of possibilities is shown schematically in Figure 5.9. Sometimes, these microstates of the whole system are referred to as arrangements or configurations.

Now, if I know the state of the atom and you have to ask binary questions, you will need to ask only a finite number of questions. This game is no different from the 20Q game which you are familiar with. Next, we move from one atom

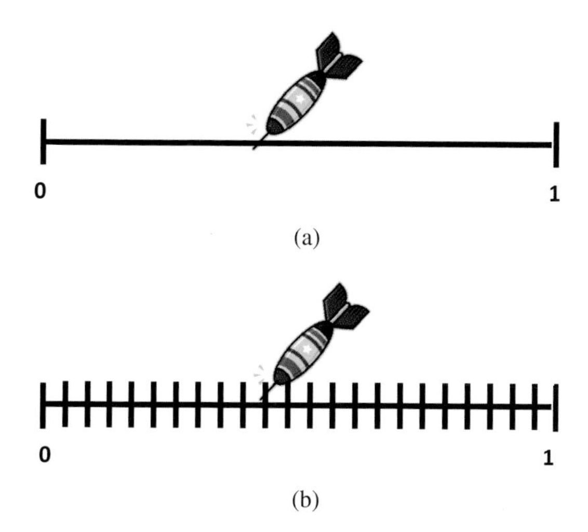


**Fig. 5.9.** The transition from the (a) continuous to the (b) discrete case.

to a huge number of atoms, say $10^{23}$ atoms, in the same box of edge d. The problem now is to find the "state" of this huge number of atoms — not the exact state, but an approximate state as is imposed by the uncertainty principle. Again, there is no difficulty in playing this game. You will need to ask many questions — far too many than you can achieve in your lifetime, or during the whole age of the universe. However, there is no difficulty *in principle* in imagining playing such a huge game. There will be a finite number of questions — finite albeit huge.

Finally, we need to introduce one principle from physics in order to reach for the entropy. The particles are indistinguishable. This means that if you exchange the locations of two atoms, you get the *same* configuration. Figure 5.10 illustrates this reduction in the number of configurations for three particles.

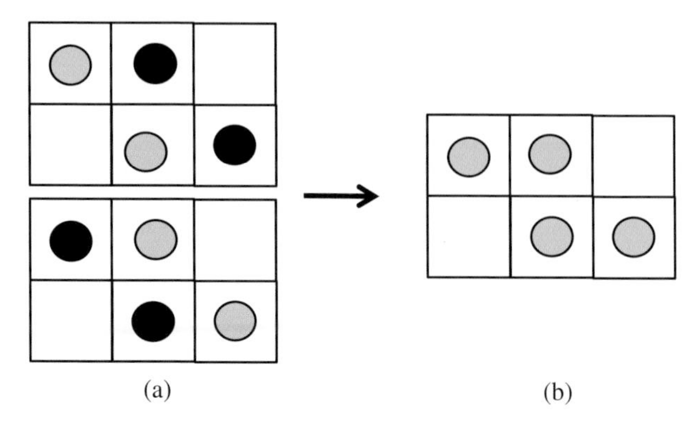


**Fig. 5.10.** The two configurations on the left (a) become one configuration when the particles are identical (b).

## 5.3 An Evolving 20Q Game

In the previous section, we introduced the 20Q game. The SMI is the average number of (smart) questions you need to ask in order to find out which "event" or which object was selected out of $n$ possible events or objects. Clearly, before you play the 20Q game you do not know which object was chosen (or where the dart hit the board). Therefore, you ask binary questions to obtain this missing information. If you must ask more questions, this means that you need more information. Therefore, the SMI is properly referred to as a *measure of information*. In all the games discussed in the previous section, the *amount of information* you lack, or the average *number of questions* you need to ask, is fixed for the particular game. It is defined by the *probability* distribution of the various outcomes, i.e. the list of all probabilities of the events associated with the specific game.

In order to understand the second law, we need to introduce a game in which the *probability distribution* itself changes with time. Hence, also, the SMI changes with time.

Consider a system having total volume $V$ divided into eight compartments (as shown in Figure 5.11), with eight marbles placed in one of these compartments. Assume, for the moment, that the marbles are distinguishable — say, having different colors, or any other labels. Now, we play the 20Q game as follows. I choose one marble and you have to find in which compartment the marble is by asking binary questions. In this, as well as in the following games, you also are given the *distribution of marbles* in the different compartments, i.e. how many marbles there are in each compartment. In the particular game shown in Figure 5.11, you know that all the eight marbles are in one compartment. This means that the *distribution* of marbles is (8, 0, 0, 0, 0, 0, 0, 0). What are the probabilities of finding the particular marble I chose in the different compartments?

This is simply obtained by dividing all the numbers in the marble distribution by 8. Hence, the probability distribution for this game is (1, 0, 0, 0, 0, 0, 0, 0). How many questions do you need to ask in order to find out in which compartment the marble I chose is? The answer is very simple. You know that all the marbles are in compartment 1. Hence, the probability that the specific marble I chose is in that compartment is 1. In other words, you *know* where the marble is. Therefore, you do not need to ask any question. In this case, the SMI is $H = \log_2 1 = 0$.

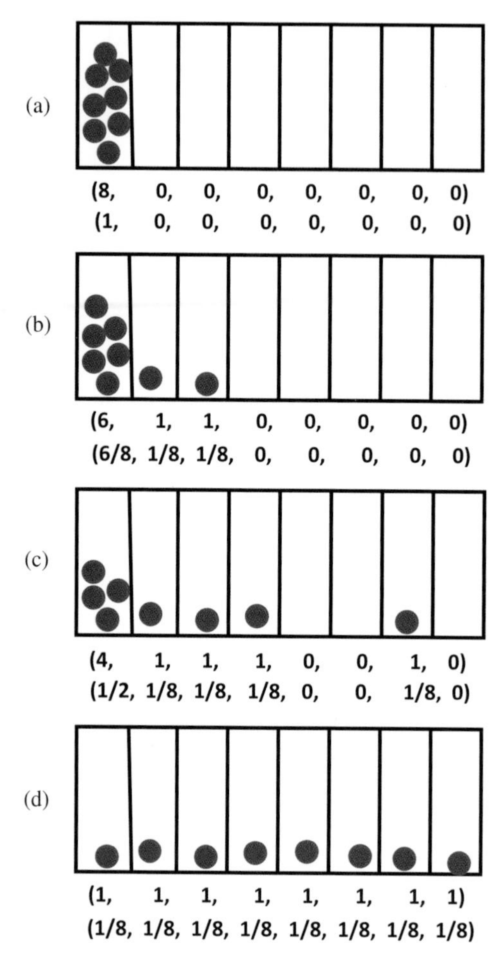


**Fig. 5.11.** A few configurations for the game with eight marbles in eight compartments.

Now, we open little holes between the compartments and start to shake the whole system. After a few minutes, some of the marbles will move to other compartments. One possible new *distribution of marbles* in the compartment is shown in Figure 5.11b. In this particular case, the distribution of marbles is (6, 1, 1, 0, 0, 0, 0, 0), and the corresponding probability distribution is $(\frac{6}{8}, \frac{1}{8}, \frac{1}{8}, 0, 0, 0, 0, 0)$. Obviously, in this game, you do not know for sure where the marble I chose is. However, you know the distribution, and this knowledge tells you that there is a relatively large probability that the marble is in compartment 1. There is a smaller probability that the marble is in the second compartment, and the same probability that it is in the third. Clearly, this knowledge can be used to reduce the number of questions you need to ask. For instance, your first question should not be: "Is the marble in compartment 6?" This will be a *waste* of a question. Also, you should not divide the total number of compartments into two groups, each of which has four compartments, and ask: "Is the marble in one of the four compartments on the right hand side?"

Thus, intuitively you feel that the *best* strategy for asking questions is to first ask: "Is the marble in compartment 1?" There is a relatively large probability (6/8) that it is there and you will end the game after one question. If you get the answer "No," then it means that the marble is in either compartment 2 or 3. Therefore, with one more question you will know where the marble is. Thus, without any calculations you feel that the average number of (smart) questions is somewhere between one and two. We say "average," assuming you play this particular game many times. Using the SMI we can calculate the average number of questions for this particular game. [For more details, see Ben-Naim (2008)].

Next, we continue to shake the system, and after a few minutes we find a new distribution of marbles in the system, such as the one shown in Figure 5.11c. Here, the distribution of marbles is $(4, 1, 1, 1, 0, 0, 1, 0)$, and the probability distribution is $(\frac{1}{2}, \frac{1}{8}, \frac{1}{8}, \frac{1}{8}, 0, 0, \frac{1}{8}, 0)$. Note that as we continue to shake the system, the number of marbles in compartment 1 *steadily* decreases, and the marbles are spread through the entire system. For some simulated experiments of this kind, see Ben-Naim (2007). While you keep this note in mind — it will be important for understanding the second law — try to answer even qualitatively the following question: "Is the game in Figure 5.11c easier or more difficult to play than the game in Figure 5.11b?" An easier game means that the missing information is smaller, and hence we need to ask fewer questions.

If you try to play this 20Q game, you will find that on average you will need about three questions (the first question should be: "Is the marble in compartment 1?" If you get a "No" answer, your next question should be: "Is the marble in either 2 or 3?" If the answer is "Yes," you will know the answer with the next question. If the answer is "No," you will also know the answer after the next question, i.e. you will find out whether the marble is in 4 or 7, and you end the game). The SMI for this game is calculated from the probability distribution $(\frac{1}{2}, \frac{1}{8}, \frac{1}{8}, \frac{1}{8}, 0, 0, \frac{1}{8}, 0)$, which is $H = 1 + 4 \times \frac{1}{8} \times 3 = 2.5$. The reason why the SMI is 2.5 and not 3 is that if you play this game many times, and always ask the first question, "Is the marble in compartment 1?", there is a relatively high probability ($\frac{1}{2}$) that you will end the game with one question. However, if the marble is not in compartment 1, you will need to ask two more questions. Thus, the average number of questions will be

$$\text{Average} = 0.5 \times 1 + 0.5 \times 3 = 2.5 \text{ questions.}$$

Before we go to the next step, note that *on average* the game becomes more difficult to play as we proceed from game 5.11a to 5.11d. The SMI steadily increases as we shake the system. If you are not convinced, try to play all the games of Fig. 5.11.

Now, we shake the system for a long time. Clearly, the initial large number of marbles in compartment 1 will be steadily reduced, and eventually the marbles will be spread into all the compartments. Typically, we will find that after a long time the marbles are spread evenly in the eight compartments. From time to time, we might see one or two marbles accumulating in one compartment, but in most of the cases we will find the distribution shown on the right hand side of Figure 5.11d, i.e. the probability distribution will be uniform: $(\frac{1}{8}, \frac{1}{8}, \frac{1}{8}, \frac{1}{8}, \frac{1}{8}, \frac{1}{8}, \frac{1}{8}, \frac{1}{8})$. You can easily calculate that in this game you will need three questions to find out where the chosen marble is.

To conclude, we started with the easiest game (zero questions to ask), and ended up with the most difficult game (three questions to ask). This trend from the easiest to the most difficult game will be maintained if we increase the number of marbles from 8 to 100, to 1000, and beyond. Not only will the trend be maintained, but we will find that deviations from this trend will be rarer as we increase the number of marbles. [For simulations of these games, see Ben-Naim (2007)]. Once you have grasped this trend, you will see that the second law is in essence nothing but the limit of this trend when we have a huge number of marbles, say $10^{23}$. We will *always* observe that the game proceeds from the easiest to the most difficult, without any noticeable deviations. Note that we mentioned the essence of the second law without using the concept of entropy. Can you guess why the distribution of marbles changed in the direction from a to d, in Figure 5.11?

## 5.4 What is Entropy?

Now that we know how to play the 20Q game (smartly), and we know that there exists a relationship between the number of objects and the number of questions we need to ask, all we have to do is to extend the game to a huge number of particles, say one Avogadro of particles, which is about $6 \times 10^{23}$.

As we have explained in the previous sections, by knowing the number of outcomes and their distribution we can calculate the SMI. A similar relationship holds for the *number of microstates* of $N$ particles and its entropy. By a *microstate* of the system we mean, classically speaking, that we know all the positions and all the velocities of all the particles in the system. In purely classical mechanics, we have infinite states for a system with any number of particles. However, quantum mechanics "reduces" this infinite number of states twice. First, by imposing the uncertainty principle, we go from infinity to some finite number of states. This reduction is achieved as we have demonstrated in Figure 5.9. We do not, and in principle we cannot, determine the *exact* location and exact velocity of each particle. Instead, we are interested in locating the state of each particle within a finite small box, as shown in Figure 5.9, but now the "box" has the size of $(\Delta l \Delta v)$ of the order of the Planck constant, where $\Delta l$ is a small length and $\Delta v$ a small range of velocities.

If we do not care about the exact location and the exact velocity of each particle, then we end up with a finite number of states for each particle. Hence also a finite number of microstates, for the entire system of $N$ particles. If you have difficulty in thinking about the abstract "space" of all the locations and all the velocities of $N$ particles, think of all the locations of being in the interval (0, 1) and reduce the number of locations from infinity to $n$, by dividing the interval (0, 1) into $n$ equal size intervals, and then disregarding the question of where exactly in each interval the particle is.

The second reduction in the number of states of the entire system is obtained by imposing the indistinguishability of the particles. This reduction is demonstrated for a small number of particles in Figure 5.10.

Once we have a finite number of states, as well as the distribution of states, we can define a 20Q game on this huge collection of states. For such a collection of states, we can also define and actually calculate the corresponding SMI. This is simply the number of binary questions we need to ask in order to determine which state was selected out of all the possible states. Of course, the SMI is a meaningful measure of the amount of information contained in the system. It does not depend on whether you actually play the 20Q game on this system.

In order to move from the SMI to the entropy, we must proceed in two steps.

First, we have to apply the SMI to the *total* number of microstates of a system with a very large number of particles (this will be important for the understanding of the second law discussed in Section 5.5).

Remember that the SMI is defined for any number of states, or outcomes of an experiment for which we know the probability distribution of the states. Examples are: (1) the toss of a coin with the distribution $(\frac{1}{2}, \frac{1}{2})$, or any other distribution; (2) the throw of a die, with the distribution $(\frac{1}{6}, \frac{1}{6}, \frac{1}{6}, \frac{1}{6}, \frac{1}{6}, \frac{1}{6})$, or any other distribution. This SMI is sometimes referred to as the entropy of the distribution. We will refrain from doing so, as this practice can lead to a great amount of confusion [see examples in Ben-Naim (2015)].

For each game, or an experiment with a finite number of outcomes, there can be many probability distributions of the outcomes. Shannon proved that for any experiment there is *one* distribution which maximizes the value of the SMI of that experiment. We have seen in Section 5.5 that in the experiment shown in Figure 5.11 there is one distribution (this is the uniform distribution shown in Figure 5.11d), which maximizes the SMI, i.e. the distribution which makes the 20Q game the most difficult to play. Figure 5.12 shows the initial locational distribution of particles in a two-compartment experiment. Figure 5.13 shows schematically how the locational distribution changes with time. In an actual experiment, the locational distribution never changes abruptly, as shown in Figure 5.12.

With this knowledge, we proceed to the next step, applying the SMI of a system having a large number of particles ($N$ of the order of $10^{23}$), at *equilibrium*. This step is important, and unfortunately it is overlooked by many authors of popular-science books.

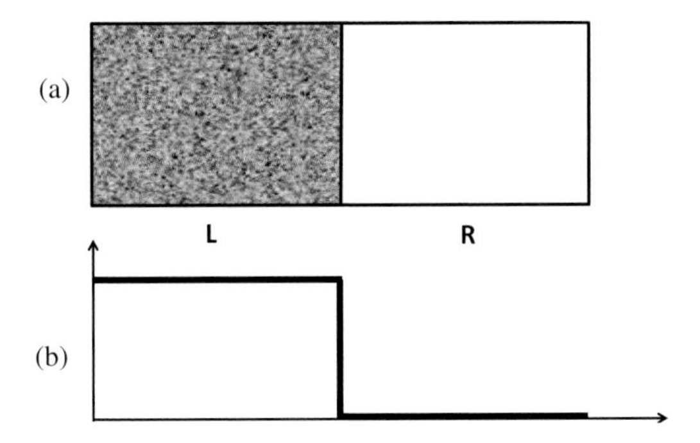


Fig. 5.12. The initial locational distribution of the particles.

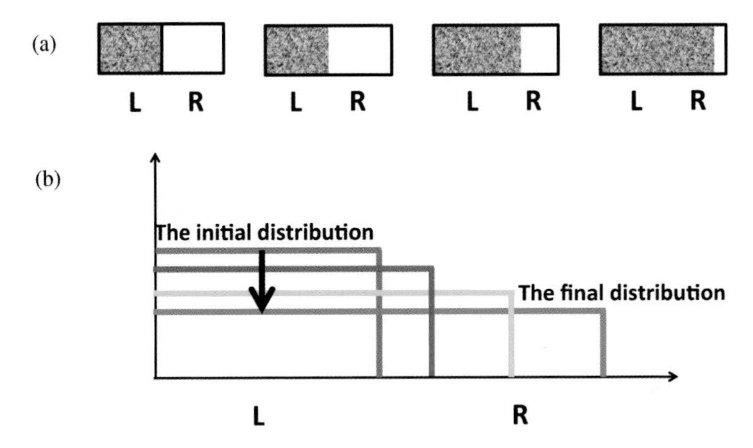


Fig. 5.13. A highly schematic changes in the locational distribution of the particle.

It turns out that for classical systems of particles, the *equilibrium* distribution of the states of the system is also the distribution which maximizes the SMI of that system. We will discuss several examples of equilibrium distribution of locations and velocities in the following sections.

One can prove that the locational distribution of particles at equilibrium is the *uniform* distribution (this is true in the absence of any external field). This means that the probability

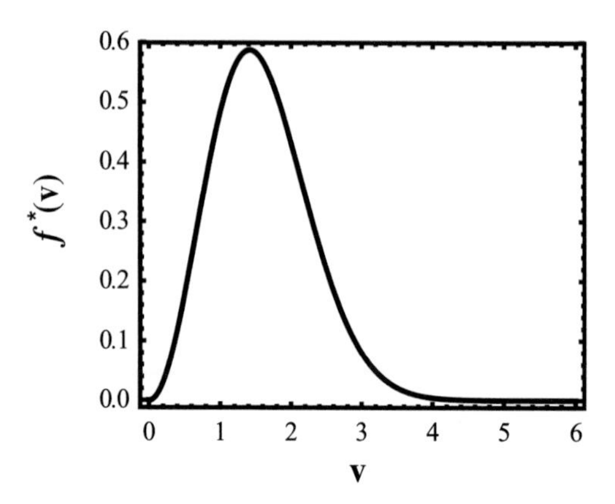


**Fig. 5.14.** The speed (or the absolute velocity) distribution of the particles.

of finding any specific particle at any specific interval $\Delta x$, or $\Delta y$ or $\Delta z$, is the same independently of the location of that interval. This result is much the same as the one we found in Figure 5.11. Only now we are considering not 8 but some $10^{23}$ particles.

Similarly, the equilibrium velocity distribution is the one shown in Figure 5.14. It is called the Maxwell–Boltzmann distribution. [For more details, see Ben-Naim (2007, 2011)].

Now, we reach the most exciting point, where we *derive* the entropy from the SMI.

First, we need to apply the SMI to the distribution of locations and velocities of all the $N$ particles in the system. Second, we have to apply the SMI for the *specific* distribution of locations and velocities at *equilibrium*.

Once you do this, you get a quantity which is almost identical with the *thermodynamic entropy*.

Incredible as it may sound, entropy, which came into being in the *field of science* in the study of the efficiency of heat engines, is related to the *average number of questions* you need to ask in order to find out in which microscopic state the system is, presuming the system is *macroscopic* and at *equilibrium*. All we need to do is to change the base of the logarithm and multiply the SMI by a constant (the Boltzmann constant) to obtain the thermodynamic entropy.

It should be said that the "microstate" of the system means the *configuration*, i.e. the locations and velocities of all the particles in the system when viewed as a *classical system*. In quantum mechanics, the "microstates" are determined by the solution to the Schrödinger equation. You do not need to know what the Schrödinger equation is. All you need to know is that if there are $W$ microstates in an *isolated* system, and if we assume that these states are equiprobable, then the entropy of the system is given by the famous Boltzmann formula $S = k_B \ln W$. This formula should remind you of the relationship between the SMI and the number of possibilities. This is strictly true for a system at constant energy $E$. When the system is not isolated, say a system at constant temperature $T$, instead of constant energy $E$, the entropy of the system is also defined by $-k_B \sum p_i \ln p_i$. Note, however that in this case $p_i$ is the probability of finding the system in a given state with energy $E_i$.

A final comment about the meaning of entropy is in order. Quite frequently, people confuse the SMI with information, which leads naturally to the association of entropy with information. From here, there is a short distance to concluding that entropy, as information, might be a subjective quantity. This kind of confusion and similar ones are discussed in Ben-Naim (2015a).

## 5.5 Does Entropy Always Increase?

Before we discuss the second law of thermodynamics, the reader should be aware of the fact that the question posed in the heading of this section is *meaningless*. If you have read popular-science books, it is most likely that you have come across statements such as "Entropy always increases." More careful writers will say: "Entropy increases most of the time." Unfortunately, both of these statements are meaningless. Entropy, in itself, does not increase or decrease, nor does it have any numerical value!

Entropy is a quantity defined for a well-defined thermodynamic system.

We have seen that the entropy of a thermodynamic system is related to the SMI of the system at equilibrium. On the other hand, one of the formulations of the second law is that if we start with an isolated system (i.e. a system having a fixed energy, fixed volume, and fixed number of particles), and we remove an internal constraint (say a partition separating two gases, as in Figure 5.1a), the entropy of the system will never decrease. It may be unchanged or it may increase.

Clausius generalized this law to claim that *the entropy of the universe always increases*.

This generalization is not warranted. First, we do not know whether the universe is finite or infinite. If it is infinite, then its SMI (and perhaps its entropy; see below) will be infinite too. In such a case, it will be meaningless to say that the entropy of the universe increases — increases beyond infinity?

Second, and a more important reservation, we do not know how to define the states of the entire universe (let alone claim that it is at equilibrium). Therefore, we cannot define the SMI of the entire universe. Hence, we cannot define the entropy of the universe.

The entropy of the universe is undeterminable experimentally, nor is it calculable theoretically! Therefore, we should refrain from talking about the entropy of the entire universe. Instead, we must limit ourselves to talking about the entropy of a well-defined thermodynamic system, say having a fixed energy, volume, and total number of particles. For such a system, it is meaningful to ask why its entropy will always increase (or stay constant) upon removal of a constraint.

The general answer to the question posed above is *probabilistic*. It consists of two parts. First, when we start with an initial state, and remove a constraint, the system will always move to a new state having a higher probability (here, by "state" we mean the thermodynamic or macroscopic state). Second, the probability of the state of the system is related to the SMI of the system. A larger probability means a larger SMI. At equilibrium the system reaches the state having the maximum probability. At this state, the value of the SMI is also maximal, and apart from a multiplicative constant this maximal SMI value is equal to the entropy of the system.

What I have written in the previous paragraph is very abstract. The interested reader can find the mathematical version of this paragraph in Ben-Naim (2008, 2011, 2015).

In this section, we will present the content of the previous paragraph by using a very simple example.

Consider a system of two compartments, as in Figure 5.1a. The entire system is isolated (i.e. there is no exchange of energy, volume, or particles with the surroundings of the system). In all of the following processes, we will start with all the $N$ particles in the left hand compartment, then remove the partition between the compartments. We will observe what happens upon the removal of the partition. How the SMI changes, how the probability of the state changes, and finally, when we get to the very large $N$, we will discuss the entropy change in this process.

In all of the following experiments, we will denote by $N$ the total number of particles. We will assume that $N$ is fixed, i.e. there are no chemical reactions, and we are not dealing with photons. The *state* of the system is characterized by the number of particles in each compartment. We denote by $n$ the number of particles in the left compartment, and by $N - n$ the number in the right compartment.

Note that since the total energy of the system is fixed, the velocity distribution in the process will not change, and as far as the locational distribution is concerned we will be interested only in the distribution regarding the question in which compartment each particle is.

Let us start with a small number of particles.

## *The case of two particles N* $= 2$

Initially, all the particles are in the left compartment. The probability distribution in this case is $(1, 0)$. The SMI of the system is 0. We know where each particle is, and therefore we do not need to ask any questions.

After removal the partition, the particles will occupy the entire system of volume $2V$. Now we can have three possible configurations: $n = 0$ with probability $\frac{1}{4}$, $n = 2$ with probability $\frac{1}{4}$, and $n = 1$ with probability $\frac{1}{2}$.

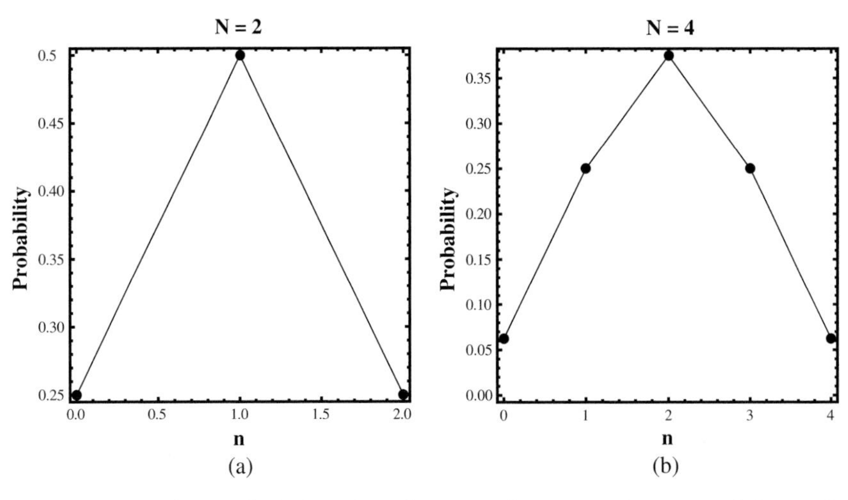


**Fig. 5.15.** The probability distributions for cases $N = 2$ and $N = 4$.

Note carefully that when we talked about marbles in different compartments, we assumed that the marbles were distinguishable. If we had two marbles instead of atoms in this particular experiment, we should count *four* configurations of the system. The corresponding probabilities are shown in Figure 5.15a. In the case of atoms, the particles are indistinguishable, and the two configurations in which there is one particle in each compartment coalesce into one configuration, i.e. $n = 1$. We write the states and their probabilities as follows:

| $n = 0$ | $n = 1$ | $n = 2$ |
|---|---|---|
| $P_N(0) = \frac{1}{4}$ | $P_N(1) = \frac{1}{2}$ | $P_N(4) = \frac{1}{4}$ |

This means that if we take many snapshots of the system, say one million snapshots, we will find the state $n = 0$ in about 25% of the snapshots, $n = 1$ in about 50%, and $n = 2$ in about 25%.

### *The case of four particles, N* $= 4$

In this case, we have four possible configurations. These configurations and their probabilities are as follows:

| $n = 0$ | $n = 1$ | $n = 2$ | $n = 3$ | $n = 4$ |
|---|---|---|---|---|
| $P_N(0) = \frac{1}{16}$ | $P_N(1) = \frac{4}{16}$ | $P_N(4) = \frac{6}{16}$ | $P_N(3) = \frac{4}{16}$ | $P_N(4) = \frac{1}{16}$ |

We see that the configuration $n = 2$ has the largest probability. Also, the probability of either $n = 0$ or $n = 4$ is relatively low. The probabilities are shown in Figure 5.15b.

### *The case of ten particles, N* $= 10$*, and more*

For $N = 10$, the distribution is shown in Figure 5.16a. We calculate the maximum at $n^* = 5$, which is $P_{10}(n^* = 5) = 0.246$. In all of the following examples, we denote by $P_N(n)$ the probability of finding $n$ out of the total $N$ particles in the left compartment. Also, we denote by $n^*$ the value for $n$ for which $P_N(n)$ is maximal.

Figure 5.16 shows the probability function $P_N(n)$, as a function of $n$ for different values of $N$. As $N$ increases, $P_N(n^*)$ *decreases*. For instance, in the case $N = 1000$, the maximal probability is. $P_{1000}(n^*) = 0.0252$. As $N$ increases, the maximal probability decreases as $N^{-1/2}$. In practice, we know that when the system reaches the state of equilibrium,

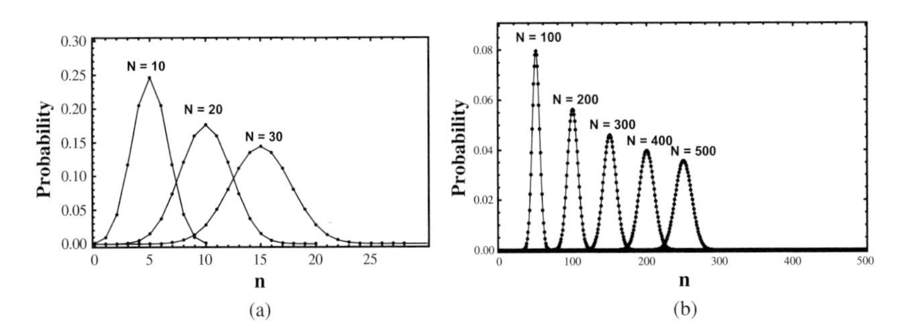


**Fig. 5.16.** The probability distributions for larger $N$.

it stays there *forever*. The reason is that the *macroscopic state of equilibrium* is not the exact state for which $n^* = \frac{N}{2}$, but it is this state along with a small neighborhood of $n^*$, say $n^* - \delta N \leq n \leq n^* + \delta N$, where $\delta$ is small enough so that no experimental measurement can detect it. The probability of finding the system in the *neighborhood* of the maximum for $N = 100$ and $\delta = 0.01$ is about 0.235. For $N = 10^{10}$ particles, we can allow deviations of 0.001% of $N$, and the probability of finding the system in the neighborhood is nearly 1.

Figure 5.17 shows the probability of finding the system having $n^* - \delta N \leq n \leq n^* + \delta N$, with a deviation of $\delta = 0.0001$ about the value of $n^*$. Plotting the probability $P_N(n^* - \delta N \leq n \leq n^* + \delta N)$ as a function of $N$ shows that this probability tends to 1 as $N$ increases. When $N$ is on the order of $10^{23}$, we can allow deviations of $\pm 0.00001\%$ of $N$, or even smaller. Yet, the probability of finding the system at or near $n^*$ will be almost 1. It is for this reason that when the system reaches $n^*$ or near $n^*$, it will stay in the vicinity of $n^*$ for *most* of the time. For $N$ on the order of $10^{23}$, "most of the time" means effectively *always*.

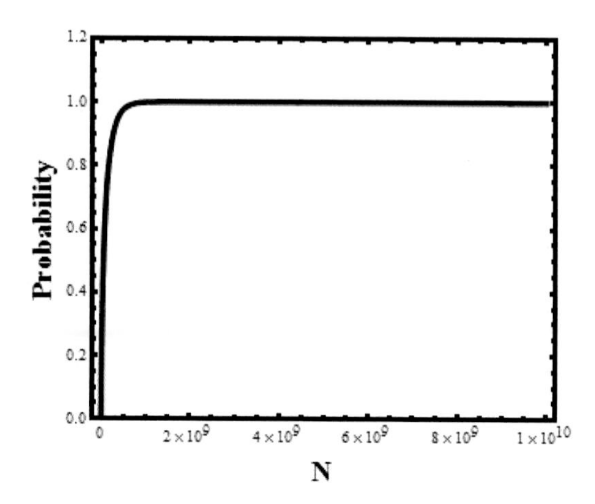


**Fig. 5.17.** The probability of finding the system having about $n = N/2$ in each compartment, as a function of $N$.

The above-mentioned specific example provides an explanation for the fact that the system will "always" change in one direction, and will "always" stay at the equilibrium state once that state is reached. The tendency toward a state of larger probability is equivalent to the statement that events that are supposed to occur more frequently will occur more frequently. This is plain common sense. The fact that we do not observe deviations from either the monotonic climbing of $n$ toward $n^*$ or staying at $n^*$ is a result of our inability to detect small changes in $n$ (or, equivalently, small changes in the SMI; see below). Note that in this section we have not said anything about the entropy changes. Before turning to calculate the entropy changes, we repeat the main conclusion of this section. For each $N$, the probability of finding a distribution $(n, N - n)$ in the two compartments $L$ and $R$ has a maximum at $n^* = \frac{N}{2}$. However, for a very large number of particles, the probability of obtaining the *exact* value of

$n^* = \frac{N}{2}$ is not very large. On the other hand, the probability of finding the system at a small vicinity of $n^* = \frac{N}{2}$ is almost 1!

When we say that the system has reached an equilibrium state, we mean that we do not *see* any changes that occur in the system. In this example, we mean changes in the *density* of the particles in the entire system. In other experiments, when there is heat exchange between two bodies we characterize the equilibrium state as the one for which the temperature is uniform throughout the system and does not change with time.

At equilibrium the macroscopic density we measure at each point in the system is constant. In the particular system we discussed above, the measurable density of the particles in the two compartments is $\rho^* \cong N/2V$. Note that fluctuations always occur. Small fluctuations occur very frequently, but they are so small that we cannot measure them. On the other hand, fluctuations that could have been measured are extremely infrequent, and practically we can say that they never occur. This conclusion is valid for very large $N$.

Let us now summarize what we have learned from the examples. For small $N$, we saw that there is a relatively high probability that there will be an equal number of particles in the two compartments. However, there is nothing irreversible in this process. The system can be found in the initial state, i.e. all the particles in the left compartment. This event has a low probability, but it is not an impossible event. When $N$ increases, we find that the probability of finding all the particles in one compartment becomes extremely small. We can observe such an event once in many ages of the universe. However, the probability of such an event is not 0. This means that we cannot claim that the process is *irreversible* in an absolute sense.

Regarding the change in the SMI, we saw that whenever we remove the partition there is a change in the SMI by the amount of $N$ bits (to determine the compartment in which a single particle is, we need one question, and hence one bit of information). This is understandable. Initially, we know that each particle is in the left compartment. After the removal of the partition, we do not know whether the particle is in the right or the left compartment. We have lost one bit of information per particle. Thus, for $N$ particles we lost $N$ bits.

So far, we have talked about the probability of the various configurations and the SMI of the particular configuration. Note also that a configuration means how many particles are in each compartment. Any configuration, say $(n, N - n)$, determines a probability distribution. Defining $p = \frac{n}{N}$ and $q = \frac{N-n}{N}$, we get the distribution $(p, q)$ for the configuration $(n, N - n)$. For each of these configurations, we calculated the probability of finding the probability distribution. We write this as $\mathrm{Pr}(p, q)$. We also defined the SMI on this distribution, which we write as $\mathrm{H}(p, q)$. It turns out that these two functions are related to each other, as we have seen in the examples above; the larger the probability Pr, the larger the value $H$. The connection between these two quantities is given in Note 4.

As we have noted earlier, the SMI is defined for any $N$, and for any specific distribution of particles $(n, N - n)$. When we remove the partition, the configuration $(n, N - n)$ will change with time. The rate of change will depend on the energy of the particles (or the temperature of the entire system), as well as on the size of the aperture between the two compartments. If we have a very small "window" between the two compartments, then the SMI will change very slowly. For any $N$, the SMI will initially increase. Remember that we started with all particles in one compartment, and the corresponding SMI was zero. When we open the window, or remove the partition, the SMI can only increase (in the language of the 20Q questions, we say that the game will always become more difficult to play). After some time, the system will attain a configuration for which the SMI has a maximal value. The configuration itself could change back to one having a lower SMI value, even to the initial configuration, which has a very low probability of occurrence.

When we want to discuss entropy, we must remember that the entropy of the system is the SMI of that system (except for a multiplicative constant), evaluated at that configuration, or at the distribution $(p, q)$ for which the SMI is maximum. In addition, if we want the entropy to obey the second law (see below), then we must apply it to a system with a very large number of particles, of the order of $10^{23}$. It is only for very large $N$ that we can say that when we reach an equilibrium state we *never* observe the reversal of the process of expansion.

## 5.6 Does the SMI Change with Time?

Before we discuss entropy and its dependence on (or independence of) time, let us go back to the experiment with marbles, as shown in Figure 5.11. In the stage shown in Figure 5.11b, we have a probability distribution $(\frac{6}{8}, \frac{1}{8}, \frac{1}{8}, 0, 0, 0, 0, 0)$. This distribution determines an SMI, i.e. $H = -\sum p_i \log p_i$, where $p_1 = \frac{6}{8}$, $p_2 = \frac{1}{8}$, $p_3 = \frac{1}{8}$, and all the other $p_i = 0$ $i = (4, 5, \ldots, 8)$. Does this SMI depend on time? Of course not! We have a fixed game as defined by the probability distribution, and this distribution determines uniquely the number of questions we need to ask in order to find the missing information, i.e. where the specific marble is.

What is the probability of finding the game shown in Figure 5.11b? You have to be careful in answering this question. Suppose I show you a die lying on the table with its upward face having four dots. I ask you what the probability is of finding the die with its upward face having four dots. The answer is 1! On the other hand, if I throw a die, and I ask you what the probability is of the outcome 4 after the die falls and settles on the table, then the correct answer is 1/6.

The distinction between the two questions I asked you about the die is important in order to understand whether entropy changes with time or not; see Section 5.7.

Let me go back to the game in Figure 5.11. Now, instead of a fixed game as shown in the figure, I shake the system for a very long time. I will take a snapshot of the compartments every second. I can also register how many times each configuration has occurred. For this kind of experiment, it is meaningful to ask: "What is the probability of finding a specific distribution of marbles?" Equivalently, we ask: "What is the probability of finding a specific *probability* distribution?"

This question can be confusing because we use here the term "probability" in two senses. One is the probability distribution, say the one corresponding to Figure 5.11b, which is $(\frac{6}{8}, \frac{1}{8}, \frac{1}{8}, 0, 0, 0, 0, 0)$. The other is the probability of finding this particular probability distribution when we shake the system for a very long time. We denote this probability by Pr, and for the specific outcomes shown in Figure 5.11 we have $\Pr(1, 0, 0, 0, 0, 0, 0, 0)$, $\Pr(\frac{6}{8}, \frac{1}{8}, \frac{1}{8}, 0, 0, 0, 0, 0)$, and $\Pr(\frac{1}{2}, \frac{1}{8}, \frac{1}{8}, \frac{1}{8}, 0, 0, \frac{1}{8}, 0)$, respectively.

These are meaningful probabilities provided we have shaken the system for a long time, and recorded many outcomes. We also know that the SMI is defined for each of these probability distributions, namely $H(1, 0, 0, 0, 0, 0, 0, 0, 0)$, $H(\frac{6}{8}, \frac{1}{8}, \frac{1}{8}, 0, 0, 0, 0, 0)$, and $H(\frac{1}{2}, \frac{1}{8}, \frac{1}{8}, \frac{1}{8}, 0, 0, \frac{1}{8}, 0)$, respectively.

Now, we can ask: "Does the SMI change with time?" Clearly, if we do not shake the system we have one probability distribution to which corresponds one 20Q game, and one value of the SMI.

The situation changes completely when we shake the system and take many snapshots. In this case, the game will change with time, and therefore the SMI will also change with time. The rate of change will depend on the way we shake the system. If we shake very gently, then the SMI will also change slowly. If we shake vigorously, then the SMI will change very fast. The point to emphasize is that the SMI, although it changes with time, is not a *function* of time. There is no function $H(t)$ that we can write down and which tells us how the SMI changes with time. The changes of the SMI with time is a result of our own decision to shake the system, and hence to change the value of the SMI at a rate we choose, corresponding to the vigor of the shaking.

Now that we understand that the value of the SMI is not a *function* of time, we proceed to the next question. Suppose we shake the system for a very long time, and we find experimentally that after a long time most of the games will have a probability distribution as in Figure 5.11d, i.e. $(\frac{1}{8}, \frac{1}{8}, \frac{1}{8}, \frac{1}{8}, \frac{1}{8}, \frac{1}{8}, \frac{1}{8}, \frac{1}{8})$. One can also prove theoretically that for this distribution the SMI has its maximum value, for this particular system of eight marbles in eight compartments. One can also prove that this probability distribution has the maximal probability of occurrence. Let us denote this maximal probability by $\Pr^{(\max)} = \Pr(\frac{1}{8}, \frac{1}{8}, \frac{1}{8}, \frac{1}{8}, \frac{1}{8}, \frac{1}{8}, \frac{1}{8}, \frac{1}{8})$, and the corresponding maximal value of the SMI by $H^{(\max)} = H(\frac{1}{8}, \frac{1}{8}, \frac{1}{8}, \frac{1}{8}, \frac{1}{8}, \frac{1}{8}, \frac{1}{8}, \frac{1}{8})$. These two quantities are related to each other.[4]

Now, for the final question. Does $(H^{(\max)})$ change with time? If you have followed me so far, you will answer this question immediately: "Of course not!" We saw that the SMI can change with time, simply because we shake the system and we cause these changes in the SMI. However, the *maximal* value of the SMI does not change with time. It is a fixed value given a fixed system. For the particular system shown in Figure 5.11, the maximal value of $H$ is simply $\log_2 8$, which is the number of (smart) questions you need to ask to find out which marble I chose in a game where all the marbles are uniformly distributed in the eight compartments.

In the previous example with marbles we found that the SMI of the game increased with time. We can say that the uncertainty associated with the location of the particular marble *increases* when we shake the system. Another way of saying the same thing is that if we play the 20Q game on this system, the game will become *more difficult* to play as we continue to shake the system. Equivalently, we need to acquire (or buy) more information in order to find out in which compartment the specific marble is located.

It is tempting to conclude from this example that the SMI always increases with time — and from this conclusion to slip into the next conclusion, that entropy also increases with time.

Such conclusions are not warranted, however. Let me give you two examples from "real" life; one in which the SMI will increase with time, and the other in which the SMI will decrease with time. In the next section, we will discuss the entropy which is basically a specific example of the SMI for a specific distribution.

### (a) ***The extreme communist society***

Suppose that in some country a new law is decreed such that all the wealth of the people in that country should be equally divided among all of them. For concreteness, suppose that there are $n$ people in the country. Suppose also that the total wealth measured in terms of the total amount of dollars in the country is $M$ dollars. In order to achieve the equality of wealth, the new law states; Every time two people, $i$ and $j$ meet, and one has $M_i$ dollars and the other $M_j$ dollars, they must redistribute the sum of their wealth, $M_i + M_j$, in such a way that after the meeting each of them gets one-half of the sum, $(M_i + M_j)/2$ (Figure 5.18). It is not difficult to

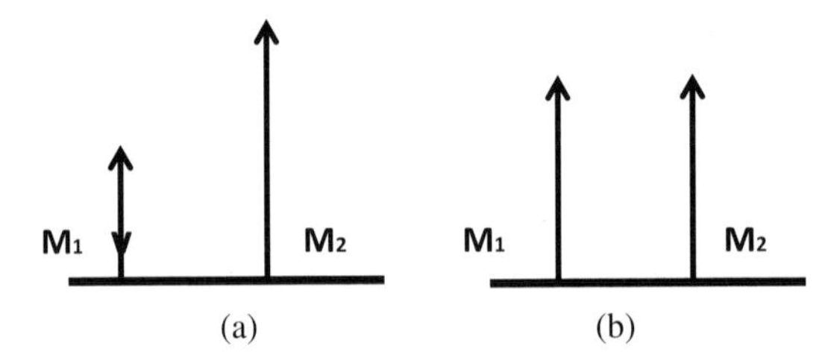


**Fig. 5.18.** The redistribution of wealth after an encounter between persons 1 and 2.

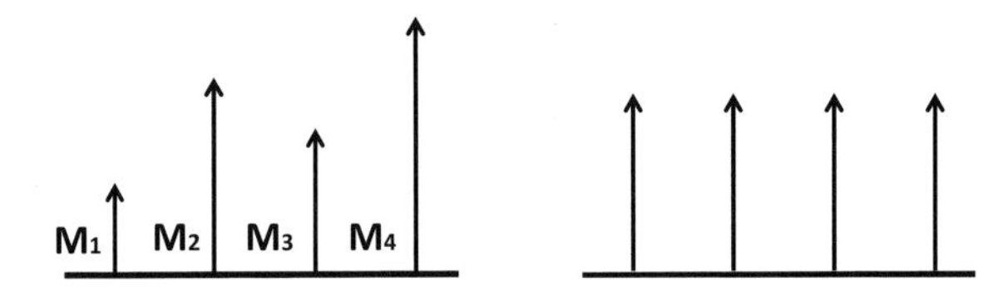


**Fig. 5.19.** The redistribution of wealth after a long time.

see that after some time (depending on the frequency of the encounters between pairs of people) the entire wealth of all the people will tend to equalize (Figure 5.19).

To define an appropriate SMI for this system, remember that the SMI is defined for any probability distribution, $(p_1, p_2, \ldots, p_n)$ such that the sum over all $p_i$ is 1. (Exercise: Can you tell for which event $p_i$ is its probability? If you have any difficulty, see the example of marbles distributed in boxes in Section 5.3.)

If person $i$ has $M_i$ dollars, we define $p_i = M_i/M$. Clearly, this is a distribution (i.e. $0 \leq p_i \leq 1$ and $\sum_i p_i = 1$) on which we can define the SMI simply by $H = -\sum p_i \log p_i$ (the log to base 2).

As we can expect, after sometime the total wealth will be divided equally between all the people. Eventually all the $p_i$

will tend to be equal to $p_i = 1/n$ (for all $i = 1, \ldots, n$). The corresponding SMI will tend to a maximum. Note that unlike the 20Q game in Figure 5.11, where the SMI could go up and down with time, when the number of marbles is very large the tendency of going up will prevail. In this country, however, once the law is decreed, the SMI associated with the distribution $p_1, \ldots, p_n$ will either remain constant or go up, and eventually reach a maximum value. Thereafter, the SMI will remain constant ($H^{\max} = \log n$).

### (b) *The extreme capitalistic society*

This example is not supposed to be a representation of a real capitalistic society. It is constructed in such a way that a distribution can be defined and the corresponding SMI will "always" decrease with time. Again, we have a country with $n$ people and a total wealth of $M$ dollars, distributed among the people. Let $M_i$ be the number of dollars belonging to person $i$; the total number of dollars is $M = \sum M_i$.

As before, we can define a probability distribution such that $p_i = M_i/M$. Clearly, $0 \leq p_i \leq 1$ and the sum of all $p_i$ is 1. In this country, people enjoy the freedom of earning as much money as they wish, and can.

One day, one person — let us call him Mr. 17 — notices that people in his country are obsessed with gambling. There are several lotteries in that country, and most if not all of the people are actively engaged in gambling. The jackpot prizes offered by the lotteries range from tens of millions of dollars to hundreds of millions of dollars. Mr. 17 also notices that the bigger the grand prize, the more people are attracted to bet, although the chances of winning are very slim. Most people do not care about the chances; each of them believes in Him, and that one day He will bestow on him or her the grand prize — no matter what the real chances of winning are.

After some deep — and hard — calculations, Mr. 17 decides to offer an exceptionally big prize of 10 billion dollars. Such a big prize is unheard of. He also tells his countrymen he chose the winning number at *random*, meaning he used a computer program which selects a random number — an integer between 1 and $10^{10^{10}}$. Note that unlike a regular lottery, a ticket sold is blank; no number is written on it. Each individual who wants to participate in the lottery will have to pay one dollar, get a blank ticket, write on it any number he or she wishes, between 1 and $10^{10^{10}}$, and hand it to Mr. 17, who will store it for safe keeping. On the first of each month, all the tickets are opened. A ticket on which the number chosen by Mr. 17 appears wins the grand prize. Note that in this lottery more than one person can win the grand prize.

Of course, no one understands what the funny number $10^{10^{10}}$ means. Whatever it means, a frenzied assault on the lottery booths ensues. After all, no one wants to miss the opportunity of winning 10 billion dollars with only a one-dollar bet. Poor people buy one ticket, choose their number, and register it with Mr. 17's lottery. Those who can afford it, buy ten tickets or even a hundred tickets, correctly believing that their chances of winning the grand prize are 10 times or 100 times bigger than for those who buy only one ticket.

Readers who have an idea what the number $10^{10^{10}}$ means should not have any difficulty in predicting how the wealth of the country will be redistributed after some time (the actual time depends on how many tickets each person buys, and on the frequency at which people buy tickets — and, of course, on how smart they are; we assume that even the smartest person will not be able to resist the temptation of winning such a huge fortune!). To the lay reader who cannot even imagine what the number $10^{10^{10}}$ means, let me tell you that if the country has a million people, and each buys a ticket every day of his or her life, the chances of winning the grand prize are nearly zero — not zero, to be sure, but very, very, very near zero.

Remember that we started with a distribution of the wealth $(M_1, M_2, \ldots, M_n)$, where $M_i$ is the number of dollars for person $i$. This distribution defined a probability distribution $(p_1, \ldots, p_n)$, where $p_i = M_i/M$ . On this probability distribution we can define the SMI, and this SMI will change with time. Before I give you the answer, try to figure out in which *direction* this SMI will change — up or down?

The answer is painfully simple. Assuming that all the people will participate in the lottery, and different people will buy a different number of tickets (remember, the more tickets you buy, the *bigger* the probability of winning; if you buy 1,000 tickets, the chances of winning are 1,000 times *bigger* than for the person who buy only one ticket).

After a short time (depending on how often the people buy tickets), all the wealth will flow in a magically single direction, into the hands of Mr. 17. Eventually, all the wealth will be concentrated in the hands of Mr. 17. The eventual probability distribution will be: all $p_i = 0$ except $p_{17} = 1$. The corresponding SMI will be zero! The change in the SMI from any initial distribution to the final distribution will be an almost steadily decreasing with time.

*Exercise.* Why did I write "almost," instead of just "steadily decreasing with time"?

*Answer.* Remember that the chances of winning are extremely small. Nevertheless, one can buy a ticket and win. This event will cause a temporary increase in the SMI (can you figure out why?). However, after a very short time the SMI will decrease almost steadily toward zero. In fact, if one person wins (in spite of the slim chances), it will *encourage* many others to invest more in the lottery, thereby accelerating the decrease in the SMI.

Now that we know the SMI can in general change either way with time, we can proceed to discuss the change in entropy with time. Before we do this, I urge the reader to try to "invent" games or events, or experiments, or even new political systems, where the SMI will go up or down with time, or perhaps oscillate with time. Also, the reader is urged to think about what it means to say that the SMI changes with time. The answer to the last question was already included in the discussion above and will be repeated in the next section.

## 5.7 Does Entropy Change with Time?

Before we discuss the answer to the question posed in the title of this section, consider the following question.

We have seen, in both Section 5.5 and Section 5.6, examples where the SMI changed with time. I trust that you could trace back the changes in the SMI to the changes in the distributions (or marbles in different cells, or of the wealth of different people). I trust also that you have noticed that the changes in the SMI are not a result of some inherent time dependence of the concept of the SMI. It changed with time because of some changes we make in a controlled experiment. Clearly, one cannot say that the SMI is a *function* of time. One must first examine whether the distribution on which the SMI is defined changes with time, and how it changes with time. The same is true of the entropy. People who talk about the ever-increasing entropy do not know what they are talking about. One must first specify the entropy of which system.

Now, consider again all the SMIs which we have studied before and which we have seen changing with time. In all of the cases, the SMI reaches some extremum value after some time (this is not necessarily always true; one can think of some oscillating process for which the SMI also oscillates). The extremum value is attained for some specific distribution. Now, I ask you: Suppose we define a system for which the distribution is the one that *maximizes* the SMI. Let us call the maximum value of the SMI $H^{\max}$. Is $H^{\max}$ a function of time?

Of course not. The quantity $H^{\max}$ is the maximum (or minimum) value of the SMI. The SMI itself might change even after reaching the maximum value, but the *maximal value* does not change with time. If you have a hard time grasping this argument, consider a simpler example. You walk along some path on a hill (Figure 5.20). At each point of time you record your height, say above sea level;

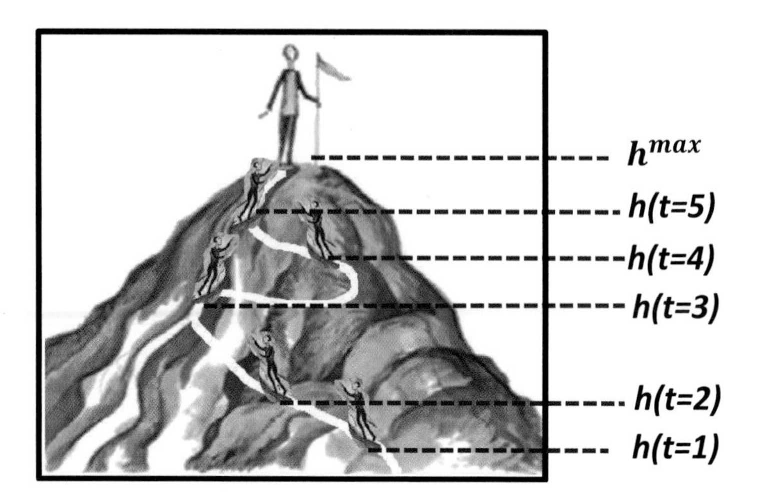


**Fig. 5.20.** The height of the person walking on the hill.

let the height at time $t$ be $h(t)$. If you continue to walk upward, the function $h(t)$ will increase with time. If you reach the summit, you record the maximal height; let us call it $h^{\max}$. Once you reach the summit, you can stay there, or you can go downhill, in which case $h(t)$ will decrease with time.

I hope you realize that no matter which way you go, and no matter at what speed you go, you will get different functions of time $h(t)$. The value $h^{\max}$ is, however, *not a function of time.* The same is true of any SMI which can change with time, but the maximal value does not change with time.

Now, we reach the crucial point. Consider the simple case of a spontaneous expansion of an ideal gas in an isolated system. Once we remove the partition, the locational distribution of the particles will change with time (here the total energy of the system is the sum of the kinetic energies of the particle; the distribution of velocities of the particles will be the same at the initial and the final state). Therefore, the locational SMI will also change with time, until it reaches the maximum value. The locational distribution that maximizes the SMI is referred to as the equilibrium distribution, and here is the bottom line: The maximal value of the SMI at *equilibrium* is related to the *entropy* of the system. Since the maximal value of the SMI is not a function of time, so is the entropy of the system.

Note carefully that the distribution of locations and velocities can deviate from the equilibrium distribution. Therefore, the SMI defined on this system may also change with time, but the entropy does not!

Now, we are ready to answer the question posed in the title of this section. The answer is: Definitely not! Of course, this question is meaningless if we do not specify the system we are dealing with. It is unfortunate that most authors of popular-science books will tell you that "entropy always increases." This is a meaningless statement. We must first describe the *system* for which we ask about the entropy. A more meaningful question is: Given a thermodynamic system, i.e. a system of $N$ (a very large number) particles in a box of volume $V$ and having a fixed energy $E$, is its entropy a function of time? The answer is: No!

## 5.8 Two Examples of Entropy Changes

Let us start with a well-defined isolated system, and remove a constraint. The entropy will increase. We can also make the change very slowly, say by opening many times the small window between the two compartments. In this case, the entropy changes will be very small. However, one cannot claim that the entropy is a function of time.

Let us go back to the evolving game in Figure 5.11. We use the same system as in Figure 5.11, but instead of marbles we have $N$ atoms of, say, argon in the system. For simplicity, suppose we start with all $N$ atoms in compartment 1. We assume that the system as a whole is isolated. It has a total fixed volume of $8V$, a total number of particles $N$, and a total energy $E$. We have chosen argon as a simple example, and we also assume that we can neglect the interactions between the atoms (ideal gas). In this case, the total energy of the system is the sum of the kinetic energies of all the atoms in the system.

If we start with all $N$ particles in one compartment as in Figure 5.21, the thermodynamic state of the system is well-defined, and we write it as the triplet of letters $(E, V, N)$. For such a system the entropy is defined — let us call it $S_{\text{initial}}$, which is determined by the variables $(E, V, N)$. If we leave the system as it is, nothing will happen; the variables $(E, V, N)$ will not change, and also the entropy of the system will not change.

We now do the analog of the shaking of the marbles as described above. All we have to do is open small windows between the compartments. Unlike in the experiment of Figure 5.11, where we had to shake the system to induce the transition from the initial states to other states, here we do not need to shake the system — the shaking is done from "inside" the system. The random kinetic energy of the atoms will cause the particles to flow between the chambers. A short time after we open the windows, only a few particles will hit the window,

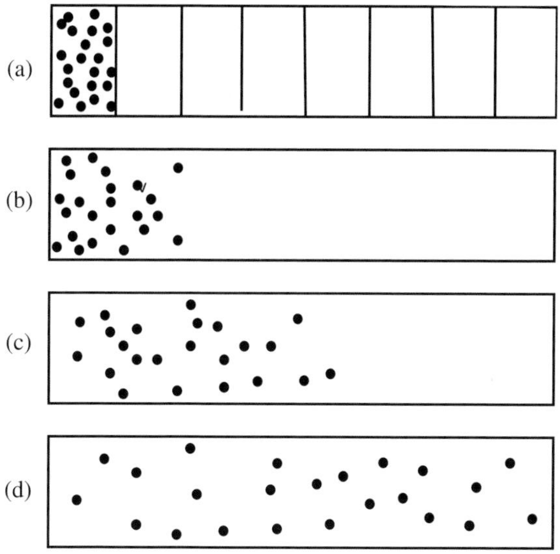

(a)

(b)

(c)

(d)

**Fig. 5.21.** An experiment of expansion of an ideal gas from the initial state (a) to the final state (d).

and enter compartment 2, then compartment 3, and so on. A few intermediate states are shown in Figure 5.21. After some time, the particles will be spread evenly over the total volume $8V$. How long it will take depends on the average speed of the molecules (i.e. on the temperature of the gas), and on the size of the windows. However, whatever the kinetic energies of the particles are, and whatever the size of the window is, we will eventually reach a new thermodynamic state, which we denote by $(E, 8V, N)$. Note that $E$ and $N$ did not change in this process. Only the accessible volume for each particle was changed. The new value of the entropy is $S_{\text{final}}$, and it is determined by the state of the system $(E, 8V, N)$.

Unlike in the experiment with the marbles for which the SMI changed while we were shaking the system, in our case the entropy of the system changes from $S_{\text{initial}}$ to $S_{\text{final}}$. This change is always positive, i.e. $S_{\text{final}} - S_{\text{initial}} > 0$. There is no function $S(t)$ which describes the change in the entropy from its *initial* value, $S_{\text{initial}}$, to its final value, $S_{\text{final}}$. Another characteristic of this experiment which is quite different than the experiment with marbles is that when the system is at equilibrium we might observe very small fluctuations from the equilibrium uniform distribution of particles. On the other hand, with the system of marbles we can observe large deviations from the most probable distribution of marbles, and for each deviation the SMI has a well-defined value, depending on the probability distribution of the marble. In fact, we can observe even the configuration when all the marbles are in *one* compartment for which the value of the SMI is zero.

The case of the gas in the compartment is very different. The value of the entropy is proportional to the *maximum* value of the SMI. The SMI will change when we open the windows, but the entropy of the system is determined by the equilibrium state corresponding to the variables $(E, 8V, N)$.

We said that there exists no function $S(t)$ which describes the change from $S_{\text{initial}}$ to the $S_{\text{final}}$. We have two states of the system: the initial state $(E, V, N)$ and the final state $(E, 8V, N)$. In this process, $E$ and $N$ remain constant. Only the volume changes, from $V$ to $8V$. The change in the SMI per particle is $\log_2 8 = 3$, i.e. three questions to determine where each particle is, exactly the same as the case of the marbles. To obtain the corresponding entropy change, all we need to do is to change the base of the logarithm and multiply by the Boltzmann constant, $k_B$.

In Figure 5.21, we show a series of possible locational distributions of particles, from the initial distribution on the right hand side of the figure to the final distribution on the left hand side. For each of these locational distributions, one can define a corresponding SMI. However, the entropy of the system is defined for two states: the initial state $(E, V, N)$ and the final state $(E, 8V, N)$. It does not change continuously from $S_{\text{initial}}$ to $S_{\text{final}}$. If you have difficulty accepting what I have just said on the abrupt entropy change, think of the volume change in the process. We start with the initial state $(E, V, N)$ having volume $V$. This is the volume accessible to all particles. When we open all the windows between the compartments, the volume accessible to all the molecules will change abruptly from $V$ to $8V$. The volume of the system is always $8V$, even if one of the compartments is temporarily not occupied by particles.

Similarly, when we open all the windows between the compartments, the entropy will change abruptly from $S_{\text{initial}}$ to $S_{\text{final}}$. As to the question of when the entropy changed from $S_{\text{initial}}$ to $S_{\text{final}}$, the answer depends on how we define the state of the equilibrium.

If we define the state of equilibrium only when the distribution of particles is uniform, and no noticeable changes can be observed, and then we can say that the entropy started at $S_{\text{initial}}$, before we opened the windows, and reached $S_{\text{final}}$ at the new equilibrium state.

On the other hand, one can decide to view each of the microscopic states of the system as an accessible state of the system. In this case, when we open the windows all the particles are still in the left compartment. However, this state is one belonging to the thermodynamic state $E, 8V, N$ (because at that point the entire volume $8V$ is accessible to all the particles). Therefore, we can say that the entropy has changed from $S_{\text{initial}}$ to $S_{\text{final}}$ at the moment we opened the windows. There are many other ways of inducing a gradual change in the entropy by opening and closing the windows at different intervals of time. We will not discuss this here. The interested reader is referred to Ben-Naim (2012).

So far, we have discussed the case of an expansion of ideal gas in an isolated system having a total energy constant. In such a process, the velocity distribution does not change. It is relatively easy to visualize how the locational distribution, and the corresponding SMI of the system, change with time. We now describe another experiment which is important in connection with the second law, where the velocity distribution changes with time.

In this example, the changes in the velocity distribution and the corresponding SMI are not so easily visualized. One needs to do some mathematics [which is available in Ben-Naim (2012)]. In the following, we will describe a process in which only the velocity distribution, but not the locational distribution.

Consider two systems of an ideal gas — say argon. One is described by the thermodynamical variables $(T_1, V, N)$ and the other by $(T_2, V, N)$. The two systems are initially isolated from each other, and are at equilibrium.

Figure 5.22 shows the velocity distributions of the two systems, say for $T_2 = 400$ K, and $T_1 = 200$ K. Note here

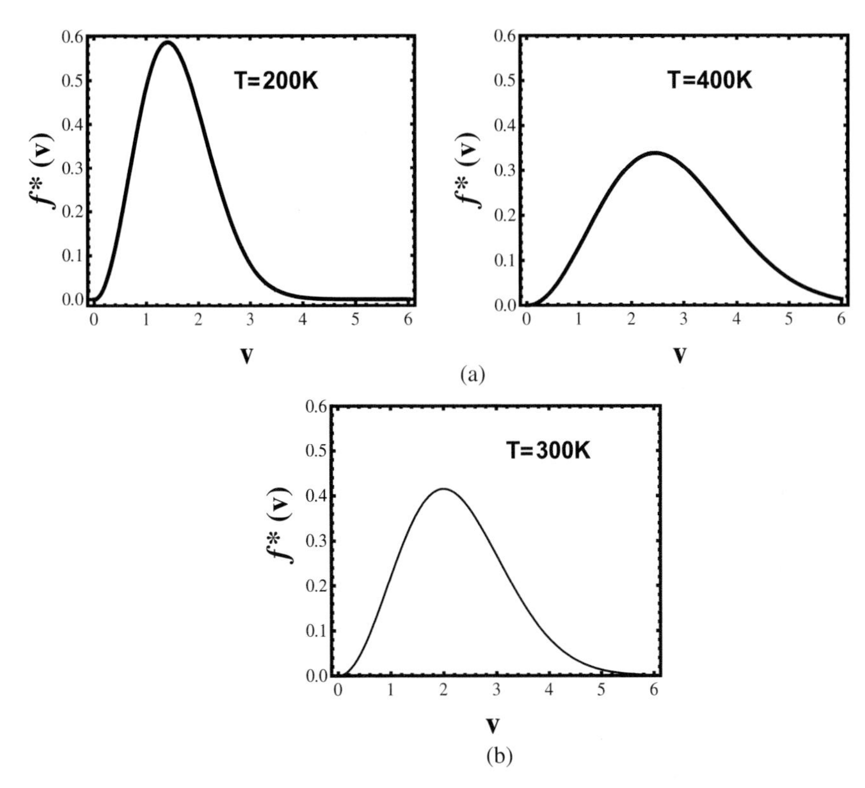


**Fig. 5.22.** The velocity distribution of the two systems in the initial state (a) and the final state (b).

that we use the term "velocity" for what is normally referred to as the *speed* of the particles.[5]

Now we bring the two systems into thermal contact. This means that heat or thermal energy can flow from one system to another. Note that the combined system is isolated. We know that heat flows from the hot body to the cold body. Also, the temperature of the hot body will gradually decrease, while the temperature of the cold body will gradually increase. At the final equilibrium state, the temperature of the two

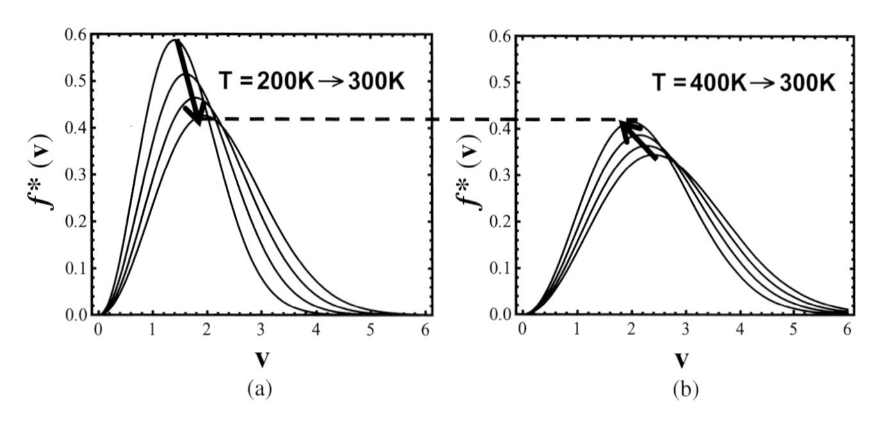


**Fig. 5.23.** The change in the velocity distribution of the two systems (a and b) from the initial to the final state.

systems will be the same, in this particular case $T = (T_1 + T_2)/2 = 300\,\mathrm{K}$.

The velocity distribution depends on the temperature. Therefore, when the temperature of the hotter system decreases, its velocity distribution becomes sharper. On the other hand, the velocity distribution of the colder system will become more widely spread. Figure 5.23 shows how the velocity distribution of the two systems will change with time. At the final equilibrium state, the velocity distributions of the two systems will be equal to each other.

Before we bring the two systems into contact, the SMI as well as the entropy of each system is well defined [for details, see Ben-Naim (2012)]. When the two systems are brought into thermal contact, the SMI of the cold system *increases* and the SMI of the hot system *decreases*. However, the SMI of the combined system always increases. This can be proven mathematically.

When we reach the final equilibrium state, the value of the SMI of the combined system is maximal. This maximal value of the SMI (up to a multiplicative constant) is equal to the entropy of the system at the final equilibrium state.

Note carefully that the SMI of the combined system changes with time once the two systems are in thermal contact. The rate of change depends on the temperatures of the system, as well as on the thermal conductivity of the surface of the systems. The entropy of the system changes abruptly from the initial value before the contact to the final value when we reach the new equilibrium state.

Thus, we can conclude that in this process the SMI increases with time (similar to the increase of SMI in the case of expansion). At equilibrium, the SMI reaches its maximal value, corresponding to the equilibrium distribution of velocities (Figure 5.22b). The entropy of the combined system is determined by the distribution of velocities at equilibrium. As such, it is not a function of time.

In the two examples discussed above, we followed separately the change in the locational distribution in the first example and the velocity distribution in the second. In more general cases, both the locational and the velocity distribution may change upon the release of a constraint. An example is the expansion of a nonideal gas from volume $V$ to volume $2V$.

In this process, we have three distributions which change: the locational, the velocity, and the interaction-energy between the particles. For each of these distributions, one can define the corresponding SMI. One can show that the total SMI of the system will always increase in this process; however, the entropy of the system changes abruptly from the initial value to its final value. [For more details, see Ben-Naim (2008, 2012).]

## 5.9 Is There a Conflict Between the Reversibility of the Equations of Motion and the Irreversibility in Thermodynamics?

The answer to this question depends on what one means by "reversibility" and "irreversibility." Ever since Boltzmann introduced the atomistic interpretation of entropy, people have been puzzled by the apparent conflict between the time-reversal symmetry of the equations of motion of the particles and the irreversibility of the second law.[6]

Of course, no one can *reverse* the time and study a process carried out by actually reversing the time itself. What one means by "time reversibility" is that the path of each particle, say the location $R$ as a function of time $t$, can occur in a reverse order. The more appropriate term is "event reversal," rather than "time reversal" of the motion of the particles.

On the other hand, we see many processes which seem to occur in one direction only — processes which are *irreversible*. For instance, a gas confined to a volume $V$ will always expand to occupy the larger volume $2V$ when we remove the partition between the two compartments, as in Figure 5.1a. We *never* see the reverse of this process occurring in our lifetime. Similarly, the process of mixing as shown in Figure 5.1b, or the heat transfer from a hot to a cold body, seems to be irreversible, i.e. we never observe the occurrence of the same process in a reverse direction.

Before we examine the question of whether there is a conflict or not, we should clarify what we mean by the "reversibility" and "irreversibility" of a process. There are at least four different senses assigned to the term "reversible process":

R1: The mechanical sense.
R2: A thermodynamic process for which the entropy change is zero.
R3: A thermodynamic process, $A \to B$, that proceeds along a dense series of equilibrium states.
R4: A thermodynamic process, $A \to B$, that can be reversed: $B \to A$.

Before we discuss the four definitions of the term "reversibility" in thermodynamics, it is instructive to see that this term could have several different meanings when used colloquially.

Suppose a person walked downhill from point A to point B. After reaching point B, you are told, the same person went back to point A. You might be wondering how this process was *reversed.*

Did the person walk backward, as would have been seen by rewinding the movie to show the person going forward and backward? This reversal of the motion would be fun to watch, but it is not a realistic process.

Another reversal process that one can imagine, but still unrealistic, is that the person went back uphill in such a way that after the completion of the cycle $A \to B \to A$ everything in the entire universe had returned to the initial state. Clearly, this is not a realistic process, but still an imaginable one.

Another reversal, now more realistic, is that the person simply went back from point A to point B along the *same path*. Although the same person retraced the *same path* going downward and upward, some changes in the person, as well as in the hill and in the atmosphere, must have occurred; for example, the soles of the person's shoes were chipped and the soil under the shoes was slightly compressed.

The fourth possibility is the simplest: the person returned to A, not necessarily along the same path, and of course changes must have occurred in both the person and the entire universe.

Bearing these examples in mind, let us move on to discuss the various meanings of "reversibility" as used in thermodynamics:

(i) The sense R1 is used in mechanics in connection with the reversibility of the equations of motion. A process is said to be reversible if by inversion of the velocities of all particles the process is reversed. This sense of the term is not usually employed in thermodynamics. It is sometimes used in statistical mechanics in connection with the apparent conflict between the apparent reversibility of the equations of motion of the atoms and molecules, and the irreversibility — in the sense discussed in (iv) below — of the spontaneous thermodynamic process.

(ii) The sense R2 is sometimes used in connection with the second law. It states that in any spontaneous process occurring in an isolated system the entropy can never decrease. A process for which the entropy increases is called an irreversible process. A process along a constant-entropy path is referred to as a *reversible* process. This nomenclature is used by Callen (1985) to distinguish between reversible and quasistatic processes. For instance, removing a partition separating two different gases will result in a spontaneous mixing, and the entropy will increase. This specific process is deemed to be irreversible, not because it cannot be *reversed*, but because it cannot be brought back to the initial state without any change in the entire universe. Strictly, reversible processes in the sense R2 rarely exist. Examples are reversible mixing of ideal gases, and change of the shape of a *macroscopic* system, say from a sphere to a cube or from a cube to a sphere (Figure 5.24). The latter is reversible in the R2 sense, provided one can neglect surface effects.

(iii) The sense R3 is most commonly used in thermodynamics. It is sometimes confused with the sense R2. The difference between the two can be clarified by the following simple example:

Consider a spontaneous process of expansion, as depicted in Figure 5.1a. Initially, all the $N$ molecules are confined to one compartment. Removal of the partition causes a spontaneous process of expansion. This is a typical irreversible

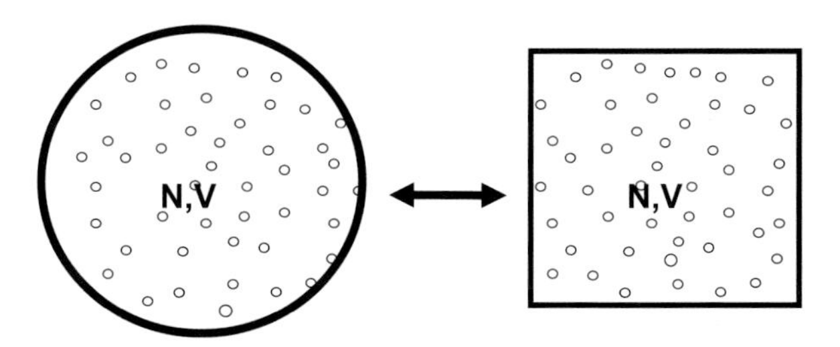


**Fig. 5.24.** A reversible process in which the entropy of the system does not change.

process in the sense discussed in (ii). The initial and final states of the system may be described by two points in the pressure–volume (PV) diagram.

Clearly, the process is not reversible in the sense R2. But it is not reversible in the sense R3 either, since we cannot trace back the path of the process in the PV diagram. This is so simply because the thermodynamic states of the system *during* the expansion process are not well defined, and therefore it is meaningless to talk about reversal along the thermodynamic path from the initial to the final state.

Suppose next that we change the weight on the piston gradually, each time reducing the weight by $\Delta M$, waiting for the system to reach an equilibrium state, then reducing the weight again by $\Delta M$, and so on. For this process, we can draw all the equilibrium points in the PV diagram. We can say that we know, say, ten points on the path from the initial to the final state, but we do not know the exact path between any two equilibrium points.

Next, we can imagine that at each step we remove only an infinitesimal weight $dM$ on the piston. In the limit, when we perform this process in infinitesimal steps, we can draw an almost continuous curve in the PV diagram, leading from the initial to the final state. Clearly, in this limiting process there is a *path* in the PV diagram leading from the initial to the final state. Therefore, it is meaningful to speak about the thermodynamic path from the initial to the final state, as well as about the reversed path from the final to the initial state.

To distinguish between this reversible process and the reversible process in the sense R2, the former is sometimes referred to as a quasistatic process. A quasistatic process is simply a process which is carried out in very small steps so that effectively the system goes through an almost continuous series of equilibrium states. Because each equilibrium state is well defined thermodynamically, it follows that the path of the process is well defined. It is therefore meaningful to talk about reversing the path, or reversing the process.

It should be stressed that a quasistatic process, i.e. a reversible process in the sense R3, does not imply that the process is reversible in the sense R2. It is therefore advisable to use two different terms to distinguish between the two.

Note also that being reversible in the sense R3 is not being reversible in the sense R1. The thermodynamic path is reversed, but the molecular trajectories are not.

(iv) The weakest form of reversibility is that a process $A \rightarrow B$ can be reversed: $B \rightarrow A$. There is no requirement that the reversal should be along the same thermodynamic path (sense R3), or that the entropy should not change (sense R2). This form is the weakest since it is difficult to find a thermodynamic process that cannot be reversed in this sense. We exclude from this discussion processes of life and death which, at least at the present level of our knowledge, seem to be completely irreversible in all the senses stated above. An example which is often cited is boiling an egg. Such a process cannot be reversed (unboil the egg?). However, in thermodynamics, we are discussing processes from one well-defined *equilibrium state* to another *equilibrium state*. It is far from clear whether an egg is in an equilibrium state either before or after the cooking. Besides, we will perhaps one day devise a process that can unboil the egg.

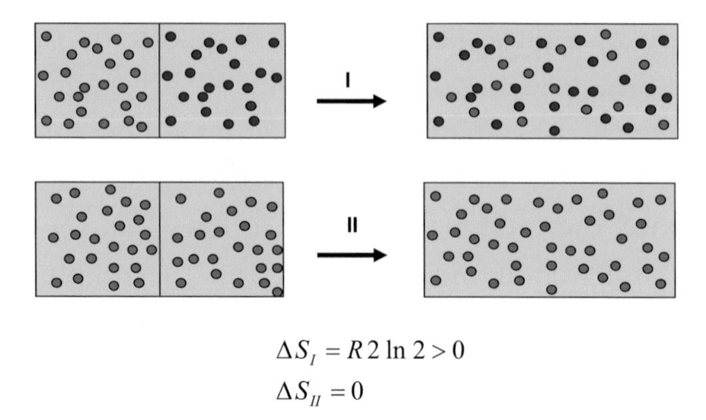


$$\Delta S_I = R2\ln 2 > 0$$
$$\Delta S_{II} = 0$$

**Fig. 5.25.** (a) An irreversible process (I) of mixing two different gases, with a positive change in the entropy; (b) a reversible process (II) of "mixing" the same gas, in which the entropy of the system does not change.

Finally, it is instructive to mention a stronger "irreversible" process that is "more absolute" than the irreversibility of the mixing process in Figure 5.25a. Gibbs, who pioneered the study of the process of mixing, compared that process with the mixing of "two gas masses of the same kind" (Figure 5.25b). He concluded that process I is irreversible (in the sense R2, i.e. $\Delta S > 0$), but can be reversed (in the sense R4). However, the mixing in process II cannot be reversed — that is "entirely impossible." This conclusion, and a similar one reached by Maxwell, seem to be correct. There is no way we can separate the two gases that are mixed in process II in Figure 5.25b. Nevertheless, reversal of process II is not only *possible*, but also trivial, and can be done effortlessly. [For further discussion of this topic, see Ben-Naim (2008).]

It is clear that when we perform a well-defined experiment, say expansion of an ideal gas from $V$ to $2V$, the *reversibility* of the motions of the particles is different from the *irreversibility* we observe for the entire system.

Suppose we perform the expansion experiment as shown in Figure 5.1a. Initially, all the particles are confined to the volume $V$. The system as a whole is isolated. This means that its total energy, volume, and number of particles are unchanged.

When we remove the partition separating the two compartments, we observe an expansion of the gas to occupy the entire volume, $2V$. The system was initially specified by the thermodynamic variables $E, V, N$. After removal of the partition, the system reaches a new thermodynamic state characterized by $E, 2V, N$, and corresponding to this change in the state of the system, the entropy will change from $S(E, V, N)$ to $S(E, 2V, N)$. We can calculate the change in the entropy for an ideal gas, and we find that $\Delta S = Nk_B \ln 2$, where $k_B$ is the Boltzmann constant.

Now, suppose we start with the final state, as shown on the right hand side of Figure 5.1a. At some point in time, we reverse all the velocities of all the atoms. Of course, in practice, we can never perform such an experiment. However, we can imagine what would happen if had we did this experiment.

We can ask the following questions regarding the outcome of such a thought experiment:

1. Will the system's particles trace back their paths and after some time we see that all the particles are confined to the left compartment?

2. Will the system's thermodynamic state reverse in the sense that it will proceed from the state $(E, 2V, N)$ to the state $(E, V, N)$?
3. Will the entropy of the gas change from $S(E, 2V, N)$ to $S(E, V, N)$, i.e. will the entropy change by the amount $-\mathrm{N}k_B\ln 2$?

Presuming that we can perform this thought experiment, we expect that all the particles will trace back their paths, and after some time we will *see* the initial state (b). In fact, the time it will take to observe this state is exactly the same as the time that passes from the time we remove the partition to the time we reverse all the velocities of all the particles. Therefore, the answer to the first question is: Yes! Regarding the second question, the answer is: Definitely no! Although we reach the initial state, this state is different from the state on the left hand side of Figure 5.1a. In other words, the thermodynamic state is not $(E, V, N)$ as in Figure 5.1a. At this state the particles are not confined to the volume $V$ as in Figure 5.1a. If we wait longer, we will find that the gas molecules will occupy the entire volume, $2V$. The thermodynamic state will be $(E, 2V, N)$.

The answer to the third question follows from the answer to the second question. The entropy will not *decrease* to the value $S(E, V, N)$. This would have been the case if the partition had also been placed at the point of time when we observed the initial state. This, of course, will not occur spontaneously.

Most people who write about the second law and admit its probabilistic nature confuse two issues.

Suppose we perform the expansion process, and after the system reaches a new equilibrium state we can ask the following two questions:

1. Will the particles of the system ever return to occupy the original volume $V$?
2. Will the entropy of the system ever decrease by the amount $R\ln 2$?

Most writers will say yes to both questions, but will add that such an event will occur with an extremely low probability.

In my opinion, the answers to these two questions are different. Yes, the particles can return to the original volume, and yes, that will occur with an extremely low probability. On the other hand, the entropy of the system will *never* get back to its original value. For this to happen, we need to assume not only that all the particles will return to the original volume $V$, but that they will stay there for a sufficiently long time to reach the original equilibrium state.

In other words, this will occur only if the partition between the two compartments returns to its original place *spontaneously*!

Thus, the answer to the second question is no. No, with no probability attached.

Note also that after we remove the partition between the two compartments, the particles *can* return to the original volume $V$ many times. You have to wait many ages of the universe for this to happen. However, the entropy of the system is determined by the variables $(E, 2V, N)$ and these will not change with time. Hence, the entropy will not change with time either.

### ***Pause and think***

If you are familiar with Maxwell's demon, who can lower the entropy of the gas by letting high-speed molecules pass one way and low-speed molecules pass the other way [for a different Maxwell's demon, see Ben-Naim (2015a)], try the following thought experiment. Consider a simpler version of the demon — call it the *patient demon* — who sits near a small window separating the two compartments. The demon waits patiently, for many ages of the universe. Once he sees that all the particles are in one compartment he closes the window, and goes to sleep. Does the entropy of the gas decrease? Is the second law violated?

To conclude this section, we can say that the conflict between the reversibility of the equations of motion and the apparent irreversibility of thermodynamics is only *apparent.* If we reverse the velocities of all the particles of a thermodynamic system, all the molecules will trace their paths and we will observe the initial states. At this moment, the molecular state of the system is different from the initial molecular state. In this new state, the locations of all the particles are the same as in the initial state, while the velocities of the particles are different (each particle will have a velocity in the reverse direction). From here on, the random motion of the particles will take over, and the law of large numbers will prevail. The system as a whole will proceed to the state having the larger probability, and this state (or rather states) will be the one characterized by the variable ($E$, $2V$, $N$), i.e. the particles will again occupy the entire volume, $2V$.

## 5.10 Does Entropy Ravage Anything?

In Section 1.10, we mentioned the expression "the ravages of time," which is frequently used by authors who write about time. As I wrote in that section, there is nothing wrong with using the expression "the ravages of time" as long as one understands that with regard to this expression there are many processes, e.g. decay, withering, and death, which we observe occurring in time. However, there is no law of nature stating that time always ravages things; time can also be constructive, as in birth, accumulation of wisdom, and many more processes occurring in time. In other words, we can use the figure of speech and say that sometimes Time ravages and sometimes Time constructs. This is tantamount to saying that sometimes we have "bad times" and sometimes we have "good times." Time, in itself, is neither good nor bad!

When one talks about *the ravages of entropy* the situation is very different. The phrase "the ravages of entropy" is always used in a negative sense. This is a result of multiple confusions:

1. Confusing entropy with time;
2. Confusing entropy with disorder;
3. Confusing the second law with a tendency toward more disorder.

The combination of all these confusions leads to the statement that the entropy of the universe always increases (with time), i.e. the disorder of the universe always increases (with time), and hence entropy will eventually lead to the thermal death of the universe (total chaos, total destruction, with everything being *ravaged* — by the entropy, of course).

In my opinion [see also Ben-Naim (2015)], entropy is even more "innocent" than time. As I mentioned above, while one can say as a figure of speech the "the ravages of time," one cannot say the same for the entropy. It is ironic that those who use the phrase "the ravages of time *and* entropy" [see Seife (2006)] actually put the blame on the entropy first, and then involve time because of the identification of the changes in entropy with the direction of the arrow of time.

The truth is that entropy does not ravage anything, nor does it construct anything — entropy simply *does not do anything*! If one uses the phrase "the ravages of entropy" as a figure of speech, it can mean only one thing: the person using this phrase has no idea what entropy is.

## 5.11 Conclusion

We have included this chapter on entropy and the second law in a book on the history of Time not because entropy or the second law has anything to do with Time, but because entropy and the second law do appear in popular-science books in connection with the so-called arrow of time (see Chapter 6).

We have seen that entropy is defined for a well-defined thermodynamic system at equilibrium. The entropy of such a system is not a function of time. Furthermore, one cannot say anything about the entropy of the universe — not in the past, not in the present, and not in the future.

Similarly, the statement that the entropy tends to increase is meaningless if one does not specify: entropy of what?

In many popular-science books, one finds statements referring to the second law as holding a special status among the laws of physics.

Here is a typical example [Greene (2004)]:

> ...it is known as the second law of thermodynamics ... Notice, though, that this is not a law in a conventional sense, since, although such events are rare and unlikely, sometimes they can go from a state of high entropy to one of lower entropy.

It is true that the second law is statistical in nature. It is not an absolute law.

All laws of physics are proclaimed to be absolute laws. As we know, there are laws or theories which were found inapplicable for some systems. In such cases, the laws have been modified, or replaced by new laws. However, there is no example of a law of physics which allows for *exceptions*. In this sense, the second law is indeed different from all other laws of physics.

The reader who encounters such a statement regarding the second law is left with the impression that somehow the second law is *weaker than or inferior to* all other laws of physics. All the laws are absolute, and there are no exceptions, but the second law does not have exceptions. The truth is, however, that the second law is *far stronger* than any other law of physics. As I have written earlier [Ben-Naim (2008a)]:

> The admitted non-absoluteness of the second law is, in fact, more absolute than any proclaimed absoluteness of any law of physics.

***Which room has higher entropy?***

# 6

# The History of the Histories of Time

In this chapter, we review some of the more recent popular-science books which discuss histories of Time, theories of Time, and interpretations of Time. While reviewing a specific book I will examine the validity of the statements made in that book, as well as provide the reader with my personal views on those statements.

I will start with Hawking's book *A Brief History of Time*, which is perhaps the first book for which the phrase "history of time" appears in the title. This review will be the most detailed one. The second book is *A Briefer History of Time*, which is a modified and shorter version of the original book by Hawking, and was published in 2004 with coauthor Leonard Mlodinow. My review of this book will be much shorter than the review of the original book. Following these two books, I will review two more books which deal with the history of Time, although the word "history" does not feature in their titles.

This chapter consists of four sections, each dealing with one book whose subject is Time. You can view these sections as extended book reviews. However, I believe there is much more than just book reviews. I hope you will also learn how to read critically any text on science written by scientists.

If you are not interested in reading the entire chapter, let me give you a brief summary of my opinion on these four books:

1. *Brief.* Over 90% of the book is irrelevant to the history of Time. This part is poorly written and most of it is incomprehensible to the lay reader. The remaining 10%, relevant to Time, is mostly meaningless and nonsensical.
2. *Briefer.* This is slightly better than *Brief*, mainly because it was cleared of most of the gibberish in *Brief.* Specifically, the 10% of the *Brief* most "relevant to time" was eliminated, rendering the book totally irrelevant to the history of Time.
3. *Eternity.* I have never heard of, seen, or read any other book with such a high density of meaningless, silly, and nonsensical statements which are repeated again and again, from here to eternity....
4. *Begin and End.* This book asks two questions. Most of the book is not about these questions. The small part of the book which is relevant to the title of the book is an over exaggeration of the *importance* and the *profound effect* of the answers to these questions for our lives. In my view, science will never have answers to these questions. And if it will have answers, they will have *zero* effect on our lives.

## 6.1 *A Brief History of Time* [Hawking (1988)]

This book was published in 1988. Since then, it has been read by millions all over the world. It is still considered a bestseller on Amazon.com. To date, over a thousand reviews have been posted on that website. Most of them (about 650) are highly positive, praising the author for his ability to explain complex ideas in simple language. A few reviewers (about 30) wrote negative reviews.

While preparing to write my present book, I read all the Amazon reviews of Hawking's book — the highly positive, the highly negative, and everything in between. My own view is that this book is poorly written, concepts of modern physics are sloppily explained (if the word "explained" can be applied at all), and above all it does not tell the "history of time" — it is at best a dull and most incomprehensible *history of science*, not a *history of Time*. In this sense, the title of the book is totally misleading.

While reading Hawking's book for the second time, I indicated all the pages of the book with three marks:

X — for pages which do not mention *Time*.
T — for pages which mention Time but are not relevant to the *history of Time*.
HT — for pages which are relevant to the *history of Time*.

The book contains 187 pages, of which 164 were marked X, 21 T, and only 2 HT. Thus, only about 1% of the book can be said to be faithful to its title. Most of the book narrates the history of science, with particular emphasis on the history of modern science.

For those who are familiar with modern science, this book will add nothing new. Most of the great questions posed at the beginning of the book are left unanswered. On the other hand, the layperson will find the book totally incomprehensible. I will quote below a few examples of paragraphs for which I have no idea what they mean. With all due respect to Hawking the scientist, I believe that Hawking the author has done a great disservice to science in writing this book.

In the acknowledgments section of the book, Hawking explains his motivation for writing this book. He writes that what led him into cosmology are the big questions, such as:

> Where did the universe come from?
> How and why did it begin?
> Will it come to an end, and if so, how?

He adds that most of the existing books do not address these questions, and that was why he attempted to write this book.

It is unfortunate that although these questions are really big ones, science does not, and perhaps cannot, provide the answers to them. In this sense, the book does not fulfill its goals.

Hawking also refers to Weinberg's book *The First Three Minutes* as a very good book. I do not share Hawking's view on Weinberg's book. My reason for this follows from what is written in Weinberg's glossary.

> Entropy: A fundamental quantity of statistical mechanics, related to the degree of disorder of a physical system. The entropy is conserved in any process in which thermal

> equilibrium is continually maintained. The second law of thermodynamics says that the total entropy never decreases in *any* reaction.

None of these sentences is correct: entropy is not related to the degree of disorder. [Examples are provided in Ben-Naim (2012).] Entropy is not conserved in *any* process in which thermal equilibrium is continually maintained. [In Ben-Naim (2015), I show examples of processes at thermal equilibrium for which entropy can either increase or decrease.] And it is not true that the total entropy (of what?) never decreases in any reaction. There are many reactions in which entropy decreases!

The quotation from Weinberg's book has been brought here not as a criticism of his book, but rather because Hawking in his book says similar things about entropy which are not true and even misleading. I will discuss examples of these later on in this section.

Carl Sagan, a great writer of popular-science books, wrote an introduction to Hawking's book. He starts with a list of great questions, similar to those listed by Hawking, but then describes more specifically what Hawking's book is all about:

> As interesting as the book's wide-ranging contents is the glimpse it provides into the workings of its author's mind. In this book are lucid revelations on the frontiers of physics, astronomy, cosmology, and courage.

It is clear from this quotation that Sagan realized that the book is *not* about the *history of Time*! As for the workings of the author's mind, the book provides unflattering, uncomplimentary glimpses.

In addition to the list of topics mentioned in this quotation, Sagan says:

> This is also a book about God...or perhaps about the absence of God. The word God fills these pages.

Clearly, Sagan must have realized what the essence of Hawking's book is: From speculations of scientists about the universe to the involvement of God in the creation of the universe — but very little about the *history of Time.*

From here on I will comment on each chapter separately.

## *Chapter 1: Our picture of the universe*

This chapter starts again with the "big questions": "the nature of time," "where did the universe come from?", "did the universe have a beginning?" and "will it ever come to an end?".

Then the author claims that some recent breakthroughs in physics "suggest answers to some of these longstanding questions." He immediately admits that some of these answers might be true, some might be ridiculous, and "only time [whatever this might be] will tell." Thus, the author starts by raising the reader's expectations of the book, but then admits that science so far has not provided any definite answer to these "big questions."

From this point on, most of the remaining parts of the book tell the history of science from Aristotle's thoughts to Ptolemy, to Copernicus, to Newton and beyond, then slip into religion, and philosophers' thinking about the beginning of time.

On page 8, the author claims:

> As we shall see, the concept of time has no meaning before the beginning of the universe.

This idea is repeatedly discussed in the book, but nowhere does the author show, as promised, that the very concept of time has no meaning before the beginning of the universe (if indeed such a beginning occurred).

The main argument is based on a Hubble observation in 1929 that distant galaxies are moving rapidly away from us, i.e. the universe is expanding. It is logical to conclude that at some earlier time all these galaxies were closer together. But then cosmologists extrapolate far back in time to what might have happened many millions of years earlier. Such an extrapolation is dangerous. Figure 6.1 shows a few galaxies drawn on the surface of a balloon. This is a common way of showing how space (represented by the surface of the balloon) is expanding and with it the distance between the galaxies is increasing. Now, if we extrapolate back in time, we might

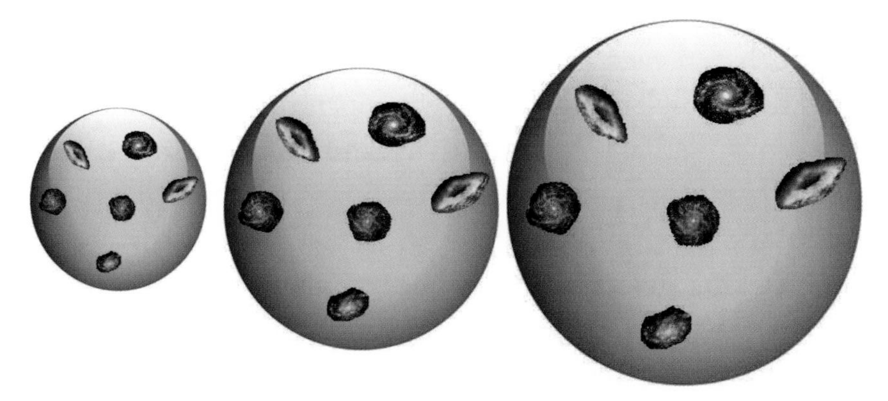

**Fig. 6.1.** The expanding universe is depicted on the surface of a sphere. As the balloon expands, all the distances between the galaxies increase.

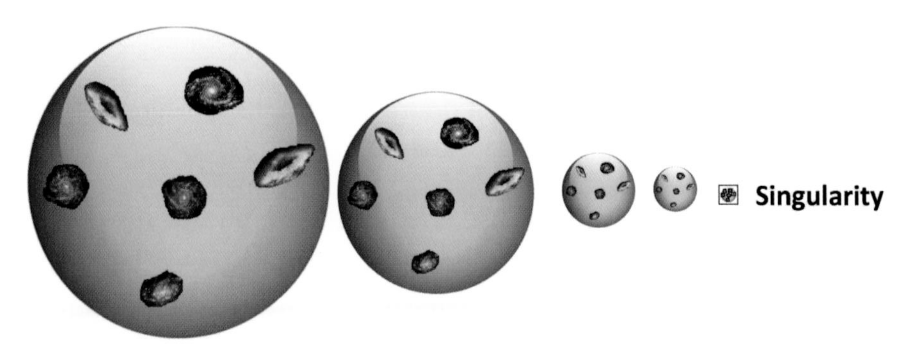


**Fig. 6.2.** Reversal of the expansion of the universe, as depicted in Figure 6.1.

simply reverse the sequence of snapshots of the universe, and "predict" the way it looked in the past (retrodict); see Figure 6.2. However, we can only make such an extrapolation up to a point. We do not know which laws of nature were operative at a very distant time when all the universe was supposed to be shrunk into a very small region. Sometimes, people even talk about a time when the entire universe shrunk to a *single point* — a point at which the density of matter and energy became *infinity*.

The truth is that we do not know how the universe would have behaved even before it shrunk to a single point. We do not know which laws of physics are operative in such extreme situations. Perhaps both relativity and quantum mechanics would break down before even reaching such a singular point of infinite density.

Hawking does admit that under such an extreme condition all the laws of science might break down, and therefore we cannot really retrodict what has happened many billions of years from now. One cannot say with *any degree of certainty* that at the Big Bang time *began*. Some physicists even claim that it is meaningless to ask "what had happened before the Big Bang," and Hawking himself, as quoted above, says:

> As we shall see, the concept of time has no meaning before the beginning of the universe.

I doubt that such a claim is meaningful. In my opinion, it is far more *meaningless* to claim that it is meaningless to ask about the time *before the beginning of time*.

When asked what God did before He created the universe, St. Augustine did not reply. His meaningful silence meant something else — He was preparing Hell for people who asked such a question. He then said that time was a property of the universe that God had created, and that time did not exist before the universe began.

In addition to Hawking's promise, "As we shall see..." (see quotation above), he now claims:

> ...if the universe is expanding, there may be physical reasons why there had to be a beginning.

And what are these *physical reasons*?

> One could still imagine that God created the universe at the instant of the Big Bang...An expanding universe does not preclude a creator, but it does place limits on when he might have carried out his job!

This can hardly be reckoned as "physical reasons" and the whole paragraph quoted above does not have a place in a popular science book!

Thus, the whole argument about the existence of the Big Bang (as well as the Big Crunch) is based on wild speculations that the laws of physics were operative for billions of years and were valid for such extreme densities. We do not know all of these. We cannot even trust any theory, not even the much-ballyhooed "theory of everything," to be applicable in such extreme situations. There is no way of either verifying or refuting such a theory.

In the next paragraphs, Hawking says that in order to discuss the question of whether the universe had a beginning or an end, we must be clear about what "a scientific theory is." This is of course true. As Hawking correctly admits:

> Any physical theory is always provisional, in the sense that it is only a hypothesis: you can never prove it.

This is *a fortiori* true when we try to use a theory of physics to either predict the distant future or retrodict the distant past of the universe. In my opinion, this point is not emphasized enough throughout the entire book, as well as in other books discussed in the following sections.

I also do not agree with the author's claim that:

> The eventual goal of science is to provide a single theory that describes the whole universe.

I am not sure that such a theory is achievable. Even if such a theory of the "whole universe" will be produced one day, it will be far from clear that such a theory will be applicable to extreme situations we have never "experienced" in our universe. We can never be sure that any theory of physics will not change with time, or even totally break down in extreme situations.

Personally, I always felt deep satisfaction at being able to solve small, perhaps even tiny, problems. The grappling with a small problem and the finding of a solution was enthralling. In my Ph.D. thesis, I tried to understand why the solvation entropy of argon in water is large and negative, compared with the solvation entropy of argon in any other liquid (see Chapter 5). This problem is not part of a grand theory of everything, and yet I felt deep satisfaction when I found an explanation for this phenomenon. What I want to say here is that I do not believe that the "goal of science" is to provide a single theory. I will be happy to have different theories for different problems.

## *Chapter 2: Space and time*

In this chapter, the author focuses on the history of *ideas* about time: from Aristotle to Newton's ideas about the absolute space and time, to Einstein's theory which abandoned that idea. In other words:

> The theory of relativity put an end to the idea of absolute time!

This idea is repeated several times throughout the chapter. The chapter ends with the following words:

> Roger Penrose and I showed that Einstein's general theory of relativity implied that the universe must have a beginning and, possibly, an end.

I doubt that one can prove such a claim from any theory of physics. In fact, Hawking admits in the next chapter that one cannot predict from general relativity anything about the beginning of the universe.

Although the pages of this chapter were marked with a T, meaning that it discusses *time*, except for the last paragraph which deals with the possible beginning of Time, there is nothing about the history of Time.

## *Chapter 3: The expanding universe*

This chapter essentially tells the same story of the expansion of the universe, the question of the beginning of time, and so on. It starts with a description of the Doppler effect, and Hubble's finding that the farther the galaxies are the faster they move away. Then there is an extensive discussion of Friedmann's models. Based on two fundamental assumptions (homogenous and isotropic of the universe), Friedmann concluded that there are three possibilities for the eventual fate of the universe. These models are poorly described in terms of the space being "bent on itself," or "space is bent the other way, like the surface of a saddle," and, finally, "space is flat (and therefore is also infinite)." I doubt that any layperson reading these lines will have any idea of the three possible Friedmann models.

This is followed by an extensive discussion of Penrose and Hawking's theory, which I found totally incomprehensible — the conclusion:

> As experimental and theoretical evidence mounted, it became more and more clear that the universe must have had a beginning in time, on the basis of Einstein's general theory of relativity. That proof showed general relativity is only an incomplete theory: it cannot tell us how the universe started off, because it predicts that all physical theories, including itself, break down at the beginning of the universe.

This conclusion is clear. In effect it invalidates most of what was explained in previous paragraphs in the same chapter. No theory can tell us "how the universe started off." No theory can be trusted in such extreme conditions which supposedly existed at the beginning of the universe (if such a beginning existed!). Thus, there is no point in telling the lay reader about all these incomplete and unreliable theories, as well as the results of such theories.

Here is a final comment on this chapter. On page 46 the author writes:

> Many people do not like the idea that time has a beginning probably because it smacks of divine intervention.

I do not agree with this conjecture. I personally do not like the idea that time has a beginning, not because it might imply "divine intervention." I do not believe that any theory of physics, not even a theory of everything, could be trusted to predict either the beginning of time or the end of time.

## *Chapter 4: The uncertainty principle*

This chapter discusses some aspects of quantum mechanics, including the two-slit experiments, the Bohr model of the atom, the wave–particle duality, and the uncertainty principle. None of these are well presented, or well explained. Nothing about time is mentioned, and of course nothing relevant to the history of Time.

The story of uncertainty in physics does not begin with the Heisenberg formulation of the principle of uncertainty in quantum mechanics.

During the 18th century, the equations of motion were formulated and reformulated by Newton, Lagrange, Hamilton, and many others. These equations became so perfect that Laplace could conclude that the universe is completely deterministic. Given all the locations and all the velocities of all the particles in the universe, Laplace claimed that this information will allow one to predict the future of the universe, i.e. the universe is completely predictable — perhaps also human behavior.

Unfortunately, even after the perfection of the classical equations of motion it is not clear whether or not such a prediction is feasible.

First, one needs an immense amount of data on all the locations and velocities of all the particles in the universe, as well as the laws of the interactions between all these particles. There is no conceivable machine which could handle this vast amount of data. (In fact, for an accurate prediction of the future, one would need infinite accuracy of the data. For any finite accuracy of the data, there will be accumulated error in the prediction — hence, loss of accuracy.) This is the reason for "inventing" of the so-called Laplace "demon," which presumably can do such a calculation. Even if such a demon exists, it will have to be able to register the data of its own particles (presuming it is made of particles), and also predict the result of its own changes and its own calculations.

Second, even if such a demon exists, and it can use all the data and calculate the future as well as the past of the universe, it is not clear how far in time one can go with such a prediction. Nothing in the classical theory of physics (or in quantum mechanics) guarantees that the law of interactions between the particles will not change over vast periods of time.

Finally, it is far from clear that the classical equations of motion (as well as quantum mechanics) are applicable to human behavior, or to any other living system.

All of these cast serious doubts on the predictability of classical physics. In fact, even before the quantum-mechanical uncertainty principle, which put Laplace's demon to sleep. In addition, quantum mechanics is inherently unpredictable. All that one can predict is the probabilities of events to either occur or not.

## *Chapter 5: Elementary particles*

This is another chapter which may be classified as "history of science," but now focusing on elementary particles starting with Democritus (who conjectured about the "grainy" nature of everything) and ending with modern theory of elementary particles (far smaller than the "indivisible" *atom*). Note that there is a distinction between "divisibility" of matter and divisibility into small parts of the same substance. The Greek idea was that if we kept cutting any piece, say of iron, we would get *pieces of iron*, but this process would come to an end when we got the smallest piece of iron — an atom, indivisible. Now we know that these atoms are in fact divisible, not to smaller pieces of *iron*, but to smaller entities which are referred to as elementary particles. How far, or rather how deep, we can dig into the atom to discover smaller and smaller particles, we can never tell.

The chapter continues with the four known forces of nature; the gravitational, the electromagnetic, the weak nuclear, and the strong nuclear force. All these are very interesting topics in the physics of elementary particles, but they do not belong to the history of time.

Toward the end of this chapter, we find a brief discussion of the C, P, T symmetries. The symmetry C means that the laws of physics are the same for particles and antiparticles, the symmetry P means that the laws are the same for any situation and its mirror image, and the symmetry T states that the laws of physics should be the same in the forward and the background direction of time. This is the only place where *time* is mentioned, not in connection with the history of time, but with the symmetry with respect to time. Finally, we read a most bizarre statement:

> Certainly, the early universe does not obey the symmetry T: as time runs forward the universe expands — if it ran backward, the universe would be contracting.

I fail to understand this sentence — why time running forward is associated with expansion of the universe, and time running backward with contraction of the universe. In my view, the universe can expand, contract, or do whatever it wishes in the same direction of time. I have no idea at all what "time runs forward" or "backward" means. I guess the author also recognized the bizarreness of such a statement, and therefore omitted this statement in the revised edition (*Briefer*) of the book (see Section 6.2).

## *Chapter 6: Black holes*

This chapter has nothing to do with the history of time, nor does it contribute anything to the history of *ideas* about time.

The term "black hole" (BH) was coined in 1969 by John Wheeler. Very little is known about BHs. One cannot carry out an experiment on BHs. They can be "felt" only through their gravitational attraction to other particles.

Most of what has been written about BHs is highly speculative. Penrose found that according to general relativity there must be a "singularity of infinite density and space–time curvature within black holes."

Whatever "infinite density" means, it is far from clear that any of the known physical theories is applicable to such singularities. In fact, the author admits:

> At such singularity the laws of science and our ability to predict the future would break down.

Not only will the prediction of the future break down, but any prediction of any property of BHs might be meaningless.

The rest of the chapter is exceptionally incomprehensible. It includes all kinds of statements for which I have no idea what they mean.

Penrose proposed:

> "The cosmic censorship hypothesis" is paraphrased as "God abhors a naked singularity."

I do, too!

On page 89:

> Anything or anyone who falls through the event horizon will soon reach the region of infinite density and the end of time.

I can imagine what will happen to anything or anyone that falls into a BH. I would certainly not dare to get close to one — but I fail to understand why that thing or person will reach the end of time, if it has any meaning at all!

## *Chapter 7: Black holes ain't so black*

This is the worst chapter of the book. First, because what it says about BHs is beyond comprehension. Second, because there is almost nothing about the history of Time. Finally, it is crammed with nonsensical statements about entropy, the entropy of BHs, and the second law. To the layperson, who does not know what entropy is, it offers confused, distorted, and incorrect descriptions of entropy. For those who are familiar with the concept of entropy, it leaves a deep impression that the author himself does not understand entropy and the second law.

If you are wondering how I got to this conclusion, here is the proof: all these nonsensical ideas were omitted in the *Briefer* edition (see Section 6.2). Here are some examples:

> The non-decreasing behavior of a black hole's area was very reminiscent of the behavior of a physical quantity called entropy, which measures the degree of disorder of a system. It is a matter of common experience that disorder will tend to increase if things are left to themselves. (One has only to stop making repairs around the house to see that!) One can create order out of disorder (for example, one can paint the house!), but that

rather requires expenditure of effort or energy and so decreases the amount of ordered energy available.

First, *entropy is not a measure of disorder of a system.* [For details, see Chapter 5 and Ben-Naim (2008, 2012, 2015)]. Such a description of entropy is very common in the literature. It should not be written by eminent scientists like Hawking.

Second, it is not a matter of common experience that *disorder will tend to increase if things are left to themselves.* I traveled abroad, left my apartment locked for a whole month, left it to itself, and no repairs were made. When I came back, I did not notice any increase in disorder. [See the figures at the beginning of Chapter 2 of my book, Ben-Naim (2015)]. Certainly, I did not notice that the "entropy of my apartment" had increased when left to itself.

The idea of leaving something to itself and it getting disordered is not only *not* true, but also irrelevant to entropy. Some books tell you that a child's room gets messier if not attended to. The fact is that the order or disorder of the room has no relevance to the entropy of the room. Therefore, such statements not only add to the confusion that already exists regarding the meaning of entropy, but also deepen the mystery enshrouding the concept of entropy.

Third, *painting the house does not create order out of disorder.*

I also have no idea what "ordered energy" is. I doubt that any lay reader would know what this means.

Finally, the opening sentence of the quoted paragraph is very misleading. The number of letters I am adding to these pages while writing this book *increases* as long as I continue to write. It is not *reminiscent* of the behavior of entropy. I can give many examples of things that increase (or do not increase) which have nothing to do with entropy. (Note that the number of letters or words I am writing never decreases. Even if I erase some letters or the entire book is burned, the number of letters I write *never decreases*. The same is true of the number of miles I walk during my lifetime. It can only increase. Even if I walk backward, the number of miles I walk will increase. It will reach a maximum when I die. All this has nothing to do with entropy, or the second Law.)

Entropy, by itself, does not increase or decrease. In some processes, in well-defined systems, entropy can either increase or decrease.

In the next paragraph, on page 102, we find:

> A precise statement of this idea is known as the second law of thermodynamics. It states that the entropy of an isolated system always increases, and that when two systems are joined together, the entropy of the combined system is greater than the sum of the entropies and the individual systems.

"This idea," presumably referring to the idea expressed in the previous paragraph, has nothing to do with the second law.

The *precise* statement of the second law is the following: In an isolated system (having a fixed energy, volume, and total number of particles), when we remove any internal constraint, say a partition between two compartments, the entropy will either increase or remain unchanged. In addition, it is not true in general that the entropy of a *combined* system is greater than the sum of the entropies of the individual systems. [For more details, see Ben-Naim (2012, 2015)].

On page 103, the author discusses the mixing of two gases. In the specific setup of the mixing experiment, the entropy does indeed increase; (see Figure 5.25a. However, the author fails to understand that the positive change in the entropy is *not* due to the increase in disorder. The author should have been aware of the fact that the change in entropy of the system is not due to the *mixing*, or to the change in disorder. [For details, see Ben-Naim (2008, 2012, 2015)].

Figure 6.3 shows mixing of two gases. Yet, the entropy of the system does not change. Figure 6.4 shows a process of *spontaneous* demixing with a *positive* change in entropy.

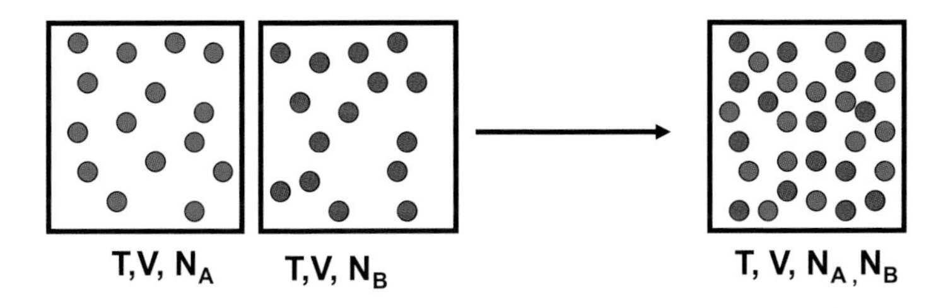


**Fig. 6.3.** Mixing two gases with no change in the entropy.

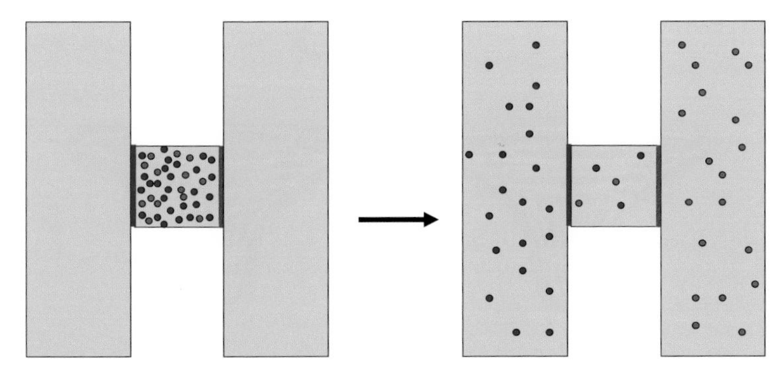

**Fig. 6.4.** Spontaneous demixing of two gases with a positive change in the entropy.

The literature is replete with formulation of the second law as a *tendency* to drive the system toward more *disorder*. There is no such law! *Disorder* does not *drive* a spontaneous process, much as a carriage does not *drive* the horses, especially a wheelless carriage. As I have explained in my more technical books, the "driving force" that underlies the second law is probability, not disorder.

Next comes another most bizarre description. What happens when we throw "some matter with a lot of entropy, such as a box of gas, down the black hole? The total entropy of matter outside the black hole would go down."

Figure 6.5 shows schematically two processes. On the left hand side, we throw a box full of gas with some entropy $S(E, V, N)$. On the right hand side, we throw a bigger box with the same energy and the same number of particles, but having twice the volume. The entropy of the second box is $S(E, 2V, N)$, which is twice as big as $S(E, V, N)$. I guess that is what the author means by "some matter with a lot of

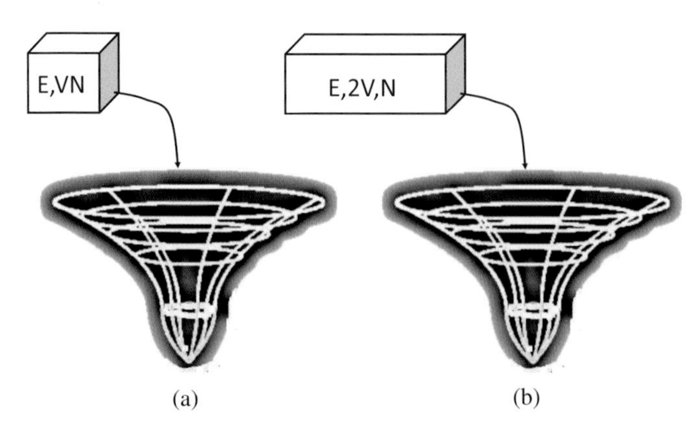


**Fig. 6.5.** (a) Throwing a box with low entropy into a black hole; (b) throwing a box with high entropy into a black hole.

entropy." Now, what will happen to the entropy "outside the black hole," "inside the black hole," or in the entire universe?

No one can answer this question. I have no idea why the "total entropy of matter *outside* the black hole would go down."

I believe that this statement is a result of the commonplace misconception about the "entropy of the universe." To the best of my knowledge, the entropy of the universe has never been defined, and I doubt if it is definable. The author admits that since we cannot "look inside the black hole, we cannot see how much entropy the matter inside it has." This is true except for what might be a slip of the tongue. We do not *see* entropy. What he means is that we cannot measure the entropy of a BH, or calculate the entropy of a BH. (Lest I be misunderstood, I should make a clarification here for the benefit of the reader who understands entropy. I said that Hawking might have had "a slip of the tongue." In fact, if you read the entire chapter you will realize that Hawking believes that entropy is a measure of disorder. Order and disorder *can* be seen. Therefore, the statement that "we cannot see how much entropy the matter inside it has" is not a slip of the tongue. It is simply a result of Hawking's misunderstanding entropy. Entropy cannot be *seen* — not in a glass of water, not in the universe, and not in a BH!)

Then the author gives his personal views on what he originally thought about the entropy of the BH, and how he has changed his mind. This entire story sounds to me like a series of absurdities. If you make an absurd assumption, then you reach an absurd conclusion. Luckily, this story, and other parts of this chapter, were deleted in the *Briefer* version.

The last part of the chapter discusses something referred to as "primordial black holes" (whatever this means):

> If the search for primordial black holes proves negative, as it seems may, it will give us important information about the very early stages of the universe.

Then comes a list of "ifs," followed by an assertion that only then could one explain the absence of an observable number of primordial BHs. I got a total blackout from reading this paragraph.

The chapter ends with an extraordinary promise:

> We shall see that although the uncertainty principle places limitations on the accuracy of all our predictions, it may at the same time remove the fundamental unpredictability that occurs at a space–time singularity.

Let us wait and *see* what the next chapters tell us about the space–time singularity.

This promise, as one can verify, is not fulfilled in any part of the book.

## Chapter 8: The origin and fate of the universe

The question posed in the first paragraph is very relevant to the history of Time:

> Does the universe in fact have a beginning or an end? And if so, what are they like?

The chapter does not tell us whether *in fact* the universe has a beginning or an end. Certainly, it does not offer any answer to the meaningless second question.

My answer to the first question is simple: No one knows if *in fact* the universe has a beginning or an end. I also believe that no one will *ever* know the answer to this question. No matter how sophisticated and honed our theories will be, we can never be sure that these theories will be applicable to such an extreme situation, in which presumably the universe *was* at the beginning or will be at the end.

I will refrain from trying to answer the second question, which I do not even understand — unless someone travels "in time" to both the beginning and the end of the universe, and tells us what those moments *are like*.

On page 117 we read:

> At the Big Bang itself, the universe is thought to have had zero size, and so to have been infinitely hot.

I can imagine what a *zero-size* universe *looks* like. But I cannot *imagine* what an *infinitely hot* zero-size universe will *feel* like. I can imagine soaring temperature, but *infinite* temperature is beyond me. All this infinite temperature is concentrated at a *zero-size universe*. I thought that temperature is a measure of the average kinetic energy of the particles. Infinite temperature must correspond to infinite average *speed* of the particles (all compressed into a zero-size space???). Is this consistent with the well-accepted fact that nothing can move with a speed larger than the *finite* speed of light? I doubt whether these questions have any meaning to cosmologists who study the conditions of the early universe. However, I do not have any doubts that such a discussion should not appear in a book addressed to laypeople.

The rest of the chapter contains many of the standard speculations about what happened after the Big Bang — after a few seconds or after hundreds, or thousands, of years. At some point, the author admits that "what happens next is not completely clear...." That is honest enough. However, it leaves the impression that what happened before that time is *completely clear*.

Then, on page 121, we find a list of unanswered questions:

1. Why was the early universe so hot?
2. Why is the universe so uniform on a large scale?
3. Why did the universe start out with so nearly the critical rate of expansion?
4. What was the origin of these density fluctuations?

I can add many more "why" questions of this kind.

Shortly after that, the author admits that general relativity on its own cannot explain these features, or answer these questions. If this is true, why raise these questions in a book on the history of Time?

As the author admits, the universe is *thought* to have been infinitely hot. This is only a *speculation* — perhaps only a mathematical result from a theory that might not be valid at the Big Bang. So what is the point of asking "Why was the early universe so hot?" if you do not know whether it was so hot?

Besides, I do not like the author invoking God so frequently to help him answer these questions:

> These laws may have originally been decreed by God.
>
> One possible answer is to say that God chose the initial configuration of the universe.

> God may know how the universe began....

While reading about this knowledge of God, I wonder whether God also knows (or knew) why Hawking wrote this book.

> The whole history of science has been the gradual realization that events do not happen in an arbitrary manner, but that they reflect a certain underlying order, which may or may not be so divinely inspired.

This is plain baloney!

The rest of the chapter is packed with meaningless, incomprehensible baloney — starting with the idea that there are "infinitely many universes," then interjecting the anthropic principle ("We see the universe the way it is because we exist"), and then the strong anthropic principle. And the most bizarre of all is the discussion about *imaginary time*. On page 138, we find:

> Only if we picture the universe in terms of imaginary time would there be singularities.

Then on page 139:

> This might suggest that the so-called imaginary time is really the real time, and what we call real time is just a figment of our imagination.
> Which is real, "real" or "imaginary time"? It is simply a matter of which is a more useful description.

No, it is not! *Real time* is real by definition, not by extent of usefulness.

Hawking should know that imaginary time has nothing to do with reality! In fact no one can *imagine* imaginary numbers. These numbers are *called* imaginary, but in fact they are unimaginable, unreal numbers.

The lay reader should be informed that imaginary numbers are used in a branch of mathematics known as complex analysis. This is a beautiful branch of mathematics, with many applications physics. In physics, especially quantum mechanics, imaginary numbers do feature quite frequently. However, whenever these numbers are used in physics, at the end, when one needs to relate a mathematical result containing an imaginary number to a real experimental result, one must cast the results in terms of *real* numbers. The reason is simple. Imaginary numbers are based on the definition of the *imaginary* square root of minus one ($\sqrt{-1}$). This number is not a *real number*, and you cannot compare any number containing $\sqrt{-1}$ with real experimental results.

Therefore, the whole discussion of "imaginary time" and the question of which is the *real* time is highly purified nonsense! Hawking should have known this.

I believe that such silly discussions may be potentially damaging to the reputation of physicists as well as to physics in general. A lay reader may easily get the false impression that physicists are a bunch of people hallucinating about the "reality" of "imaginary" numbers.

Here are two exercises with an important moral:

*Exercise 1*. By throwing a die many times, I found the relative frequencies of the appearance of each of the faces: 1, 2, 3, 4, 5, 6. These frequencies may be viewed as the "experimental" probabilities of the various outcomes. Denote by $p_i$ the probability of finding the outcome $i$. I tell you that if I square the probability $p_4$ (i.e. take $p_4^2$), and add to the result the number $\frac{1}{16}$, I get $\frac{1}{8}$. What is the probability $p_4$?

This is an easy problem. You can see the solution in Note 7. There are two solutions to this problem. Which solution would you accept as *real*?

A more difficult exercise is the following:

*Exercise 2.* I tell you that I took the cube of $p_4$ (i.e. $p_4^3$) and added to it $\frac{1}{64}$, and I got $\frac{1}{32}$. What is the probability $p_4$?

There are three solutions to this mathematical problem (see Note 7). Which one would you choose as the real solution?

Once we are done with the exercises, let us get back to the "deep" ideas about *real* time and *imaginary* time. We can thank God (or perhaps I should thank Mlodinow) that all these pieces of nonsense were omitted in *Briefer* (see Section 6.2).

So far, the author has not shown what he promised at the end of Chapter 7.

Before I go to the end of the chapter, I would like to ask the reader, including the author, to explain to me what the following sentence means.

> The poor astronaut who falls into a black hole will still come to a sticky end; only if he lived in imaginary time would he encounter no singularities.

Does this mean that if the astronaut lives in *real* time, he will encounter singularities?

I am clueless as to what this sentence means, assuming that it means anything at all. I would be grateful to anyone who could offer help in clarifying this matter. I also promise to include any clarification in a second edition of *Briefest*, even though it is not relevant to the history of Time (or perhaps it is relevant?).

The chapter ends up involving God yet again, and again, and again....

> The idea that space and time may form a closed surface without boundary also has profound implications for the role of God in the affairs of the universe.

Really?

> Most people have come to believe that God allows the universe to evolve according to a set of laws and does not intervene in the universe to break these laws.

I believe that God would never have allowed such nonsensical statements made on His behalf....

> It would still be up to God to wind up the clockwork and choose how to start it off. So long as the universe had a beginning, we could suppose it has a creator. But if the universe is really completely self-contained, having no boundary or edge, it would have neither beginning nor end: it would simply be. What place, then, for a creator?

That is a lot of baloney. Hawking should *know* that God created the universe in six days. He did not "wind up the clockwork." He simply did not have time to do that. It was already Friday evening, and He had to do the Kidush. Then He had to rest for the whole of the Shabbat.

Such a paragraph should be placed (by the creator, of course) in a book on theology, not in a popular-science book.

## *Chapter 9: The arrow of time*

The pages of this whole chapter were marked with a T and some with an HT (see the introduction to this section). This means that this chapter is *most relevant* to the history of Time, more than any chapter of this book. In spite of this, the lay reader might find it very unfortunate that almost all of this *relevant* stuff was omitted in *Briefer*. Such an omission renders *Briefer* almost empty of any topics relevant to the history of Time (see Section 6.2).

To the lay reader I would say that it was a very judicious decision to eliminate almost the entire chapter in *Briefer*. The reason, I guess, is that the author (or authors) realized that what is written in this chapter is even more nonsensical than what was written in the previous chapter and therefore, they fortunately, decided to eliminate it.

I will not bother the reader with a detailed discussion and criticism of all the nonsense written in this chapter. To do this, I would have to copy almost the whole chapter — which would make this book far longer than I planned. Instead, I will quote some outstanding juicy sentences and comment on them, allowing the reader to ponder on them. Have fun!

The first paragraph is about the transition from "absolute time" (before relativity) to "relative time" (after relativity). This transition has been discussed more than once is this book. Also, the three symmetries C, P, and T were discussed in Chapter 5. I do not see any reason to repeat that here.

Page 144:

> This says that in any closed system disorder, or entropy, always increases *with time*. In other words, it is a form of Murphy's law: Things always tend to go wrong!

Entropy is not disorder, and Murphy's law has nothing to do with the second law.

Page 145:

> There are at least three different arrows of time.

These are:

1. The thermodynamic arrow of time;
2. The psychological arrow of time;
3. The cosmological arrow of time.

All of these *arrows* are in the index of this book *Brief*, but mysteriously disappeared from the index of *Briefer*. Why? In *Brief*, almost eight pages are devoted to the arrows of time, but not even one sentence on these arrows appears in *Briefer*. Why?

In my opinion, all of these arrows are figments of the author's imagination. There is no thermodynamic arrow of time. Thermodynamics does not have any arrows!

The psychological arrow of time is, as it says, a *psychological* arrow. As discussed in Chapter 1, we feel that time sometimes runs very fast and sometimes terribly slow. Sometimes we have a good time, and sometimes we have a bad time. Therefore, it is clear that there are many, perhaps infinite, psychological arrows of time, not one! Each person, and perhaps each animal, has many psychological arrows of time, not one! All of these have nothing to do with physics, and certainly, not with the history of time.

Finally, if you wish, you can define the cosmological arrow of time. But one should not connect this arrow of time with the "direction of time in which the universe is expanding rather than contracting."

Such a connection leads some people to conclude that during the expansion of the universe time changes in the direction of the arrow of time, but when it contracts (if it will contract) it will change in the opposite direction to the arrow of time (whatever this might mean).

Page 145:

> I shall argue that the psychological arrow is determined by the thermodynamic arrow, and that these two arrows necessarily always point in the same direction.

This is, of course, a preposterous idea. As far as I understand it, there are infinite psychological arrows of time (see above), but there are *zero* thermodynamic arrows of time! How Hawking managed to *determine* infinite arrows from zero arrows is a remarkable achievement.

And what is the thermodynamic arrow of time? According to the author (page 145):

> The direction of time in which disorder or entropy increases.

Then, the author asks:

> Why does disorder increase in the same direction of time as that in which the universe expands?

Unfortunately, these are nothing but mumbo jumbo. Disorder *does not* increase with time! Entropy is not disorder!

If you buy a lottery ticket every day, you will most likely lose money. If the lottery is such that there is *one* winning number and $10^{10^{30}}$ losing numbers, I can guarantee that your wealth will steadily dwindle with time. This does not mean that the lottery has an arrow of time.

And to the question "Why does disorder increase in the same direction of time...?" I can offer an immediate answer. It is exactly the same answer you would have given to the following question: "Why does beauty increase in the same direction of time?"

A better question would be: "Why does silliness increase in the same direction of time?" Any answer by readers would be appreciated.

The most laughable paragraph appears on page 146. It is high-grade nonsense.

> Suppose, however, that *God decided* that the universe should finish up in a state of high order but that it didn't matter what state it started in. At early times the universe would probably be in a disordered state. This would mean that disorder would *decrease* with time. You would see broken cups gathering themselves together and jumping back into the table. However, any human beings who were observing the cups would be living in a universe in which disorder decreased with time. I shall argue that such beings would have a psychological arrow of time that was backward. That is, they would remember events in the future, and not remember things in the past. When the cup was broken, they would remember it being on the table, but when it was on the floor, they would not remember it being on the floor.

After all this gibberish, the author admits: "It is rather difficult to talk about human memory because we don't know how the brain works in details." This is the only sober sentence on this page, which effectively invalidates all that is said in the above-quoted paragraph.

Having done with the psychological arrow of time, the author moves on to discuss "the psychological arrow of time for computers." Whatever this might mean (nothing!), the reason cited by the author for introducing this fictitious arrow of time is somewhere between nonsensical and insane.

> I think it is reasonable to assume that the arrow for computers is the same as that for humans.

It is not only "reasonable to assume" that the arrow for computers is the same as that for us humans. They are in fact *exactly* the same — both are fictitious arrows, and both are *reasonably the same* as the arrow of the iPhone, the arrow of the television, the arrow of a cockroach which I just saw racing against my psychological arrow of time.

Here are some more puzzles:

> Our subjective sense of direction of time, the psychological arrow of time, is therefore determined within our brain by the thermodynamic arrow of time. Just as a computer, we must remember things in the order in which entropy increases. This makes the second law of thermodynamics almost trivial. Disorder increases with time because we measure time in the direction in which disorder increases. You can't have a safer bet than that!

Indeed, the psychological arrow of time is subjective. However, we have no idea how our brain feels the arrow of time. We also have no idea whether thermodynamics, specifically entropy and the second law, has any relevance to the brain. Therefore, some people might "remember things" in the order in which entropy increases, some others remember things in the order in which entropy decreases, and some do not remember whether entropy "increases" or "decreases"! As for myself, I admit that I lost all "my entropy" upon reading that quotation.

I fully agree that the second law is *trivial*, and I have shown this in previous chapters, and in more detail, in Ben-Naim (2008, 2012, 2015). In fact, I showed that the second law is not a law of physics but a law of common sense. The second law has *nothing* to do with decreasing order, or with the (nonexistent) "direction in which disorder increases"!

And, of course, "you can't have a safer bet" on the merits of this entire paragraph. I just finished *ordering* my desk, and I found that it took one *positive* hour. Did I defy the second law as stated by Hawking?

As the question the author asks — "Why should the thermodynamic arrow of time exist at all?" — I would simply answer that the thermodynamic arrow of time does not exist. Therefore, this is an empty question! It is like asking: "Why should the beauty arrow of time exist at all?"

I definitely disagree with the author that the last question is equivalent to:

> Why should the universe be in a state of high order at the end of time, the end we call the past?

The answer to the second question is exactly the same as the answer to the following question: "Why should the universe be in a high state of silliness at one end of time?" Any answer is acceptable!

On page 150, the author admits that he had made a mistake (he first believed that disorder would decrease when the universe recollapsed). Then he indirectly compliments himself by saying that it is *better* and *less confusing* if one admits in print that he was wrong. Unfortunately, the author fails to admit that almost all of what he says in this chapter is either wrong or not even wrong. The fact that almost all of the content of this chapter was omitted from *Briefer* is tantamount to indirect and implicit admittance of that.

I cannot resist a few more sentences that are exceptionally meaningless:

> ...a strong thermodynamic arrow is necessary for intelligent life to operate.
> ...intelligent life could not exist in a contracting phase of the universe.
> ...intelligent beings can exist only in the expanding phase.

Aha, now I know that since I have a *very weak* arrow of Time, I cannot operate either in the expanding universe or in the contracting universe. What a tragic revelation.

Here, the author goes from silly to sillier to the silliest statements. I would welcome any suggestions by readers which could decipher for me these quoted statements.

## Chapter 10: The unification of physics

At the end of Chapter 9, the author promises that in the next chapter he will try to explain how people are trying to fit together the partial theories...to form a complete unified theory that would cover everything in the universe.

As always, such promises are not fulfilled. Chapter 10 does not attempt to find a unified theory of physics, sometimes referred to as the theory of everything (TOE).

I found most of this chapter incomprehensible. I also dislike again involving God in answering questions that physics cannot provide answers to.

This chapter, even if comprehensible, is totally irrelevant to the history of Time. I will therefore skip it and discuss the next chapter.

## *Chapter 11: Conclusion*

> We find ourselves in a bewildering world. We want to make sense of what we see around us and to ask: What is the nature of the universe? What is our place in it and where did it and we come from? Why is it the way it is?

These are beautiful questions. Unfortunately, no one can answer such questions within physics, and no (meaningful) answers are offered in this book. Besides, what have all these questions (and answers) got to do with the history of Time? On page 173, the author summarizes that part of the theory which is relevant to the history of Time:

> According to the general theory of relativity, there must have been a state of infinite density in the past, the big bang, which would have been an effective beginning of time. Similarly, if the whole universe recollapsed, there must be another state of infinite density in the future, the big crunch, which would be an end of time.

Here, we have the *entire* history of time from the beginning, at the Big Bang, to the end, at the Big Crunch. Nothing is between these ends. The lay reader should be aware of the fact that even these unique "events" associated with time are based on a mathematical theory which encounters bizarre singularities at both ends of time. The relevance of these mathematical results to reality is at best doubtful.

Having said that, I cannot resist commenting on the insistence of the author on invoking God again, even in the last sentence of the book:

> ...why it is that we and the universe exist. If we find the answer to that, it would be the ultimate triumph of human reason — for then we would know the mind of God."

This is really a great idea: To know the mind of God!

It is unfortunate that this paragraph was repeated in *Briefer*.

## 6.2 *A Briefer History of Time* [Hawking and Mlodinow (2005)]

The motivation for writing this book (*Briefer*) is provided in the foreword:

> One repeated request has been for a new version, one that maintains the essence of *Brief History* yet explains the most important concepts in a clearer, more leisurely manner.

I found that in this sentence the authors admit that the original version of the book (*Brief*) was not clear enough.

There is no doubt that this is a better-written book, giving better explanations and almost devoid of all the nonsense that fills *Brief*.

It is ironic, however, that this reduction in size, from *Brief* to *Briefer*, was achieved by eliminating almost all the topics which are in principle relevant to the history of time, thus rendering the title of the book totally inappropriate. A better title could be "A Briefer History of Science" or, better yet, "A Briefer History of Modern Physics."

If you compare the indexes of *Briefer* and *Brief*, you will find that all the following topics have disappeared:

Arrow of Time
Entropy and Black hole entropy
Cosmological arrow of Time
Psychological arrow of Time
Thermodynamic arrow of Time
Thermodynamics, second law of
Time, arrow of
Time, imaginary

The authors do not explain why they omitted all these topics from *Briefer*. This is a serious flaw of *Briefer*. In *Brief*, Hawking admits several times that he was wrong — and he praises himself for the courage to do so. Unfortunately, nowhere in either *Brief* or *Briefer* does he admit that he was wrong in almost everything he said about the arrow of time, the involvement of entropy and the second law in the history of Time. The fact that these topics have been left out is a silent yet screaming admission.

Thus, the improvement of the present book was achieved by rendering it devoid of anything which is pertinent to the history of Time. Instead, one can find in the index new entries, such as "antigravity," "dark matter," and "dark energy," which have nothing to do with the history of Time.

As I did with *Brief*, while reading *Briefer* for the second time I marked all the pages of the book with HT (where the history of time is mentioned), T (where time is discussed), and X (where time is not discussed).

Out of a total of 153 pages (compared with 187 in *Brief*), 1 page was marked HT (2 in *Brief*), 15 were marked T (21 in *Brief*), and 137 were marked X.

In the rest of this section, I will briefly comment on some statements of *Briefer* which do not appear in *Brief.*

Chapter 1, titled "Thinking About the Universe," is almost the same as Chapter 1 of *Brief.* The questions most relevant to the history of Time are:

1. What is the nature of time?
2. Will it ever come to an end?
3. Can we go backward in time?

At the end of the chapter, the authors admit that they do not have answers to these questions.

We are informed that recent breakthroughs in physics *suggest* answers to some of these questions, but it is far from clear whether these answers will someday seem obvious, or perhaps ridiculous. "Only time [whatever this may be] will tell."

In my opinion, the authors could end the whole book at this point. They not only do not have answers to the questions raised regarding time, but they also admit that they do not have a clear-cut idea of what time is! The saying "Only time will tell" is, of course, a figure of speech. Time does not tell us anything. Similarly, when they ask "Will it ever come to an end?" the "it" is "time." Rephrasing this question, we get "Will time ever come to an end?". Again a figure of speech — time does not go anywhere, and will not come to anywhere, and certainly not to *anywhen*.

Chapters 2 and 3 contain similar topics to Chapter 1 of *Brief.* Basically, this is a very short history of science — not a history of Time.

Chapter 4, on Newton's universe, discusses Newton's laws, particularly Newton's law of gravity. This law states: If the mass of a body is doubled, so is the gravitational force that it exerts. In other words, this law states that the *gravitational attraction* is proportional to the mass of the *body* being *attracted*, as well as to the mass of the body which *attracts.* An illustration is shown in Figure 6.6. The illustration on page 21 of *Briefer* shows the author sitting in an automated wheelchair, presumably being *doubly attracted* by two *bodies* of the same lady. I found this illustration to be in poor taste, to say the least. It certainly *does not* illustrate Newton's law of gravity!

Chapter 5 narrates a short history of the ether, the classical experiment of Michelson and Morley in 1887, where they measured the speed of light in different directions. They found that the speed of light is the same in whichever

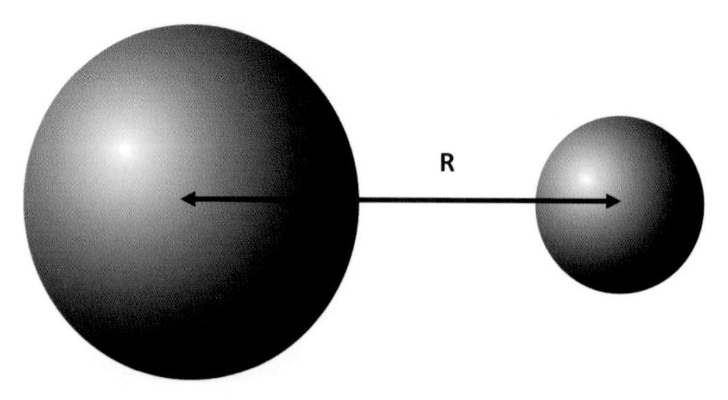


**Fig. 6.6.** The gravitational force is proportional to the product of the masses of the two particles.

direction it travels. This, and other experiments, led Einstein to conclude that time is not absolute — as was thought by scientists and philosophers from Aristotle to Newton. Furthermore, time is not completely *separated* from and *independent* of space, but is combined with it to form the object called space–time.

The authors do not explain what "separated" and "independent" mean in the context of time and space. These are meaningful, technical words which should be explained to the lay reader. Instead, the authors show an illustration on page 34 which is supposed to "explain" the four dimensions of space–time, but in fact it illustrates only three-dimensional space, and again, in my opinion, this *illustration illustrates* bad taste in *illustration*!

It is true that any event can be described by a point (or region) in space, and a point (or a period) in time. However, I do not agree with the following statement (page 35):

> ...in relativity, there is no real distinction between the space and the time coordinates, just as there is no real difference between any two space coordinates. (See Note 1.)

If this were true, why not use four coordinates — a, b, c, d — to describe an event without mentioning the words "time" and "space"?

Chapter 6, on curved space, is a short description of the general theory of relativity. The authors discuss the effect of gravity on time, but I found the whole explanation unclear, and while some of the illustrations are beautiful, they do not add much to clarify the discussed issues. Finally, the chapter ends with a long sentence:

> As we shall see, the old idea of an essentially unchanging universe that could have existed forever, was replaced by the notion of a dynamic, expanding universe that seemed to have begun a finite time ago, and which might end at a finite time in the future.

I doubt that what the authors promise to show is actually shown in the rest of the book. Personally, I do not believe we can ever know whether time began some time ago, or will come to an end sometime in the future.

Chapter 7 discusses the expanding universe. I found that the topics discussed in this chapter (Doppler effect, Hubble findings) are much better explained than in *Brief*. Then, there is a long discussion of Friedman's models and Friedman's assumptions about the universe, which I doubt can be understood by the lay reader. Again, the illustrations are beautiful, but add very little to the understanding of the text. In particular, the illustration on page 55, on the "blackbody spectrum," shows two balls, presumably at different temperatures, radiating some waves, but it does not explain anything. This chapter concludes with the statement about the ever-expanding universe, and about the ever-*increasing* rate of the expansion. From this the authors conclude:

> Time will go on forever, at least for those prudent enough not to fall into a black hole.

I doubt that such a conclusion is warranted. Chapter 8 focuses mainly on the Big Bang and on the beginning of time. I have no idea what is meant by:

> At the Big Bang itself, the universe is thought to have been infinitely hot.

To me, temperature is associated with the kinetic energy of particles. Therefore, infinite temperature would mean infinite energy too, and perhaps also infinite velocities of all the particles in the universe condensed at a single point (infinite density). No explanation whatsoever is provided in this book.

On page 69, the authors specifically explain:

> ...temperature is simply a measure of the average energy — or speed of the particles.

They use this interpretation of temperature to explain the cooling effect of the universe when it expands. Unfortunately, they forget that the same interpretation of temperature when applied to the singularity of the Big Bang leads to an absurd result of the infinite temperature implying infinite speed!

This absurd result is only one example of many mathematical results that are obtained from the general theory of relativity. The authors do not emphasize enough that mathematical results of a theory are not always relevant to the real world. There are many theories involving algebraic equations which have more than one solution [examples in Ben-Naim (1992)]. Some of these solutions are physical absurdities (such as negative probability or imaginary time).

One must be careful enough to choose the mathematical solution which is physically meaningful. It might also be the case that all mathematical results are not physically meaningful. In this case, one must examine the theory itself, as well as its applicability in a particular extreme situation. I suspect that the absurd result obtained for the Big Bang would turn on a red light in the mind of a theoretician about the validity of the theory he or she is using.

It is unfortunate that such results are propagated in popular-science books which are addressed to laypeople having no tools to assess their validity.

On page 76, the authors discuss the events that occur at different periods of time after the Big Bang. They admit that "what happens next is not completely clear. . . ." As if what happened before, including the Big Bang itself, is "completely clear."

Toward the end of Chapter 8, the authors admit that the theory of general relativity is an incomplete theory because it cannot tell us how the universe started off.

This chapter also includes a discussion of black holes, and all kinds of fanciful stories about the fate of an astronaut who approaches a black hole, and even an illustration of such an astronaut near the surface of a black hole (page 81). One important missing topic here is the fanciful story of black hole entropy which filled several pages of *Brief*. Somehow, with a stroke of a magician's wand, all these stories about black hole entropy disappeared. Not only "black hole entropy" disappeared, but also the almighty "entropy" itself, and the second law and their relationship with the arrow of time. All have disappeared, leaving no traces in *Briefer*. Why? There is a good reason for that, but unfortunately the authors do not offer any explanation.

Chapters 9 and 10 contain a few new topics, which were not included in *Brief.* However, the whole of Chapter 9 of *Brief,* on the arrow of time, is missing in *Briefer.* A whole chapter which is relevant to the history of Time is missing. It was removed without a single word of explanation, leaving the careful reader (of both *Brief* and *Briefer*) wondering. Why? My answer has been provided in the previous section.

Instead of all the fanciful discussions about the various arrows of time, the authors discuss other fanciful topics such as traveling backward in time.

Then we find a long discussion of Gödel's solution to Einstein's equations, and whether the whole universe is, or is not, rotating (whatever this means). I failed to understand what the message of this chapter is. The many illustrations do not help the reader make sense of the text. Perhaps one particular illustration is clear. On page 105, we see the authors sitting on a time machine. So now the reader has an idea of what a time machine looks like.

The chapter ends with a completely incomprehensible paragraph:

> We can avoid these problems if we adopt what we might call the chronology protection conjecture. This says that the laws of physics conspire to prevent macroscopic bodies from carrying information into the past.

This sounds as if the authors conspired to confuse the reader to such an extent that he or she, or I, would be prevented from understanding, or rather acquiring information from, this book.

Finally, they conclude that the possibility of time travel remains open.

> But don't bet on it. Your opponent might have the unfair advantage of knowing the future.

I have no idea what the authors are conspiring to tell us in this whole paragraph.

Chapter 11, on the forces of nature and the unification of physics, is full of wild speculations about the ultimate fundamental particles which are the most elementary building blocks of the universe. Nothing in this discussion is relevant to the history of Time — in particular, the concluding sentences of this chapter referring to the inability of the laws of physics to predict human behavior from mathematical equations. A fact every student already knows!

The concluding Chapter 12 is very similar to the concluding chapter of *Brief.* One outstanding figure, on page 139, shows the universe carried by three turtles on one side and a funnellike figure on the other side. The caption says: "Ancient and modern views of the universe." Now I know what the modern view of the universe looks like — something like a funnel.

The concluding chapter ends with the same sentence as in *Brief.* Again, leaving the reader the full freedom to figure out whatever he or she might mean by:

> …for then we would know the mind of God.

## 6.3 *From Eternity to Here: The Quest for the Ultimate Theory of Time* [Carroll (2010)]

This book was published after both *Brief* (1988) and *Briefer* (2005). In the book's bibliography we find a reference to *Brief*, but not to *Briefer*. I would venture a speculation that Carroll did not read *Briefer*. Had he read that book, he would have noticed that many of the topics discussed in *Brief* are missing from *Briefer* . I believe that had Carroll understood why these topics were removed (though very relevant to the history of ideas about time), he would not have written this book.

I experienced *déjà vu* while reading Carroll's book, as it contains all the nonsensical stuff that is in *Brief*, but excluded in *Briefer*. Additionally, it contains many new, but no less nonsensical, topics presented in a detailed, overly repetitious and exaggerated manner. All the topics which were meticulously removed from *Brief* came crawling back in, and came back with a vengeance in *Eternity*.

In my opinion, this book is below par and is probably one of the most nonsensical popular science books I have ever seen or read. To begin with, I have no idea what the title *From Eternity to Here* means. I understand that it is wordplay, with the 1953 movie *From Here to Eternity* as the author's inspiration, but for a popular-science book which deals with the quest for the ultimate theory of time, I feel that this title is meaningless.

Perhaps the original title, *From Here to Eternity*, would be far more appropriate for this book. The reason is that the book is full of nonsensical statements, which are repeated again and again, then again and again with little variations on the same theme, then goes on and on...from here to eternity!

The subtitle of the book is also meaningless and misleading. The author should know that no theory in physics can be the *ultimate* theory. One can never be sure that an existing theory in physics will not be modified as a result of new experiments or theories. Therefore, it is an exercise in futility to search for the ultimate theory in physics. It is preposterous to search for the *ultimate* theory of time where no theory of time exists.

As stated in the prologue:

> This book is about the nature of time, the beginning of the universe, and the underlying structure of the physical reality.

It is not about a theory of time, and certainly not about the (meaningless) ultimate theory of time.

Two *grand* questions are posed in the prologue:

> Where did time and space come from?
> How is the future different from the past?

And then, the author promises:

By the end of this book we will have defined *time* very precisely, in ways applicable to all fields.

Reading through the whole book (some parts I had to read twice), I did not come across any precise *definition* of time — or, of course, see any answers to the above quoted questions.

Let us go through those parts of the book which discuss time, and see what the author believes is a precise definition of time.

Page 2:

> The most mysterious thing about time is that it has a direction: the past is different from the future. That is the arrow of time.

First, I do not see any mystery in the fact that time has direction. As I explained in Chapter 1, "time's direction" is an acceptable figure of speech. Events unfold in a sequence which we ascribe to a sequence of points *in time*. This is how we measure time, and how we use it in any other experiments carried out in a sequence of points *in time*.

> A major theme of this book is that the arrow of time exists because the universe evolves in a certain way.

This is an empty statement, not only because we do not know whether an arrow of time exists or not, but more importantly, even if the universe would have evolved in any other way, this would have no effect on the "existence" of the arrow of time.

> The reason why time has a direction is because the universe is full of irreversible processes — things that happen in one direction of time, but never the other.

This is followed by a connection of irreversible processes with the second law, and a connection of the second law with "something called *entropy*, which measures 'disorderliness' of an object," and that "Entropy has a stubborn tendency to increase, or at least stay constant, as time passes."

Can you please explain — entropy of what has a stubborn tendency to increase?

All this nonsense appears in *Brief*. We have discussed these aspects of entropy and the second law in previous sections of this book.

> But there is one absolutely crucial ingredient that hasn't received enough attention: If everything in the universe evolves towards increasing disorder, it must have started out in an exquisitely ordered arrangement. This whole chain of logic, purporting to explain why you can't turn an omelet into an egg, apparently rests on a deep assumption about the very beginning of the universe: It was in a state of very low entropy, very high order.

The crucial mistake of the author is the (unfounded) assumption "that everything in the universe evolves towards increasing disorder…." If that assumption were true, then of course the universe must have started in a highly organized ordered state. Unfortunately, the assumption itself is not true. In fact, it is far from clear what "order" or "disorder" of the universe means.

Consider this valid logical deduction: Everything in the universe evolves toward increasing ugliness. It follows that the early universe started in a very beautiful state. This conclusion is as valid as the author's conclusion quoted above. No one can prove that my statement is wrong. The same is true of the author's statement.

Thus, what Carroll calls "a deep assumption about the very beginning of the universe" is at best a very deep speculation and at worst a meaningless speculation. And, of course, the connection of all this "disorderliness" with entropy is pure nonsense. Note that we are still in the prologue of the book. We will read about all these senseless ideas, and repeated at that, throughout the book. You will recognize the metaphor supposedly representing the second law: "An untended room tends to get messier over time." Not my room! Sometimes, when I tend it, it gets messy, and sometimes it gets tidy — but whenever it is untended, it does not change.

> The arrow of time is the reason why time seems to flow around us, or why (if you prefer) we seem to move through time. It's why we remember the past, but not the future. It's why we evolve and metabolize and eventually die. It's why we believe in cause and effect, and is crucial to our notions of free will.
> And it's all because of the Big Bang.

It is correct that time *seems* to flow around us, and I do prefer to say that we seem to move through time. But what has that got to do with "why we remember the past, but not the future"? We remember the past because some events which occurred in the past are recorded in our brains. This is the same reason we know the history of the past because some events are recorded in our brains, in books, on tapes, or on CDs. No events from the future are recorded in our brains, and therefore we cannot retrieve nonexistent information. And all this has nothing to do with the Big Bang!

By the way, I do not remember what my great grandfather did on July 1, 1900 (the past). But I do remember that I have an appointment with the dentist tomorrow (the future). Does this also follow from the Big Bang?

The reader should be informed that the Big Bang is a highly speculative idea based on extrapolating back into the past, using theories that might not be applicable to such an extreme state of the universe.

We do not know whether the Big Bang occurred or not. (In my opinion, it most likely did not). However, attributing all the processes listed in the quotation above to the Big Bang, is both irresponsible and highly misleading, especially for the lay reader.

Reading the above quotation led me to wonder: "What if the Big Bang never existed?" Would we remember the future but not the past? Would effects precede causes? Would we die first, and then get younger and younger in the direction of the arrow of time? And what would happen to "free will"?

Here is another beautiful piece of nonsense:

> The mystery of the arrow of time comes down to this: Why were there conditions in the early universe set up in a particular way, in a configuration of low entropy that enabled all of the interesting and irreversible processes to come? That's the question this book sets out to address. Unfortunately, no one yet knows the right answer.

*I know, I know, I know the right answer!* Here it is: The conditions in the early universe were set up in such a perfect configuration that it was love at first sight when the entropy and the energy met...lo and behold, the fruit of that love was born: the arrow of time.

The entropy of the universe is not definable.[8] Therefore, it is meaningless to talk about the "low entropy" of the early universe. Thus, the question "the book sets out to address" is also meaningless. So why write over 400 pages of a book on a meaningless question to which "no one yet knows the right answer"?

Yet, surprisingly, the author asks the question again and again, invoking the same idea about the "entropy of the early universe," which is referred to as the "past hypothesis," which is anything but an empty hypothesis.

Fortunately, there is one nice anecdote, which in my opinion mitigates the reading of this book.

> "An old professor who listened to the author's lecture did not find the talk convincing. He sent the author an email commenting that 'to suggest that the law of physics depends on the magnitude of the entropy of the universe is *sheer nonsense.*' And Carroll's statement that the Second Law owes its existence to cosmology is one of the dumbest [*sic*] remarks I heard in any of our physics colloquia."

After quoting the professor's email, the author says: "I hope he read this book."

I, Arieh Ben-Naim, do hereby declare that I read the entire book, some parts more than once, and I totally agree with the professor's comment, which in my opinion is an understatement.

In the last paragraphs of the prologue, the author outlines the plan for the book, and concludes with an honest statement:

> All of which is unapologetically speculative but worth taking seriously.

So, let us see what is "worth taking seriously."

### *Chapter 1: The past is present memory*

To the best of my knowledge, the past is all that has happened in the past. It has nothing to do with what is in the memory. Perhaps I am missing something deep in this title. Let us see if the content of the chapter sheds light on its title.

To the question "What do we mean by time?" the author provides three answers. But before doing so he invokes entropy, which is completely out of context: "You don't walk down the street and bump into some entropy." Indeed, I never bumped into some entropy, but what has that got to do with the question about the meaning of time? Here are the answers to that question:

1. Time labels moments in the universe.
2. Time measures the duration elapsed between events.
3. Time is a medium through which we move.

I believe that everyone will agree with these characterizations of time, colloquially speaking. Surprisingly, most of the rest of the chapter is devoted to showing that these three ideas about time need not be related to one another. I fully agree with the author about the "flow of time." I have discussed this in Chapter 1. Unfortunately, I could not see anything in this chapter which would justify its title: "The past is present memory." I still do not understand what the title means.

## *Chapter 2: The heavy hand of entropy*

Can anyone explain to me what this title means?

The chapter opens with a most impressive sentence:

> Forget about spaceships, rocket guns, clashes with extraterrestrial civilizations. If you want to tell a story that powerfully evokes the feeling of being in an alien environment, you have to reverse the direction of time.

I am left perplexed. I have no idea what this opening sentence means. I am very eager to "tell a story..." Unfortunately, I do not know how to "reverse the direction of time." Can anyone help? Then we find:

> The consistent increase of the entropy throughout the universe, which defined the arrow of time.

This, and similar meaningless statements, reverberate through the entire book. Let me reiterate: The entropy of the universe is not definable. Therefore, it is meaningless to say that the entropy of the universe consistently increases. And, it is *a fortiori* meaningless to claim that this entropy, or any other entropy, *defines* the arrow of time.

And what is entropy?

> Specifically, it measures how disorderly the system is.

It is not! I will suggest to the author to have a quick look at the two figures in the opening chapter (on entropy) of my recent book. *Information, Entropy, Life and the Universe* [Ben-Naim (2015)].

Then, there is the totally misleading example of a collection of papers stacked on top of one another, having *low entropy*, and the same collection of papers scattered haphazardly, which is supposed to have *high entropy*. I can guarantee that if you measure the entropy of the two collections of papers shown in Figure 6.7, you will find that the entropy of the ordered stack is *exactly* the same as the entropy of the scattered papers.

Since the actual measurement or calculation of the entropy of this collection of papers is not easy, let me rephrase what I said above in a slightly different way: *The difference between the entropies of the two states of the collection of papers on the left and on the right hand side of the figure is zero!*

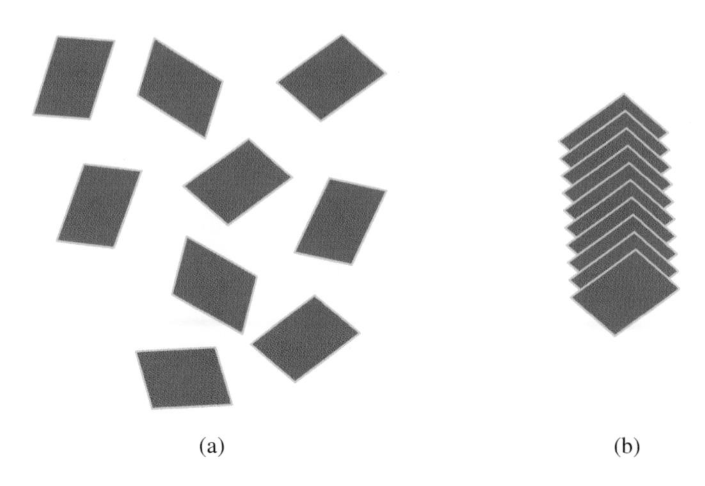


**Fig. 6.7.** The scattered and stacked collections of papers.

Now, if you still have doubts about the entropies of the two configurations in Figure 6.7, do the following "more realistic" exercise.

Suppose you have $N$ pages scattered, but at a very large distance from one another (Figure 6.8). And suppose that initially the pages have slightly different temperatures, say $T_1, T_2, \ldots, T_N$. At very large distances these pages do not *interact* with one another. Now, we put all the pages in one perfectly ordered stack. The pages interact with one another, and after a short time the temperature of all the pages will be the same [assuming identical pages, the final temperature will be the average $(T_1, T_2 + \cdots T_N)/N$].

Would you expect the entropy of the ordered stack of pages to be larger than, smaller than, or equal to that of the disordered stack? The answer is provided at the end of this section.

If you are not convinced, let me give you one more example, similar to the one shown in Figure 6.7, but

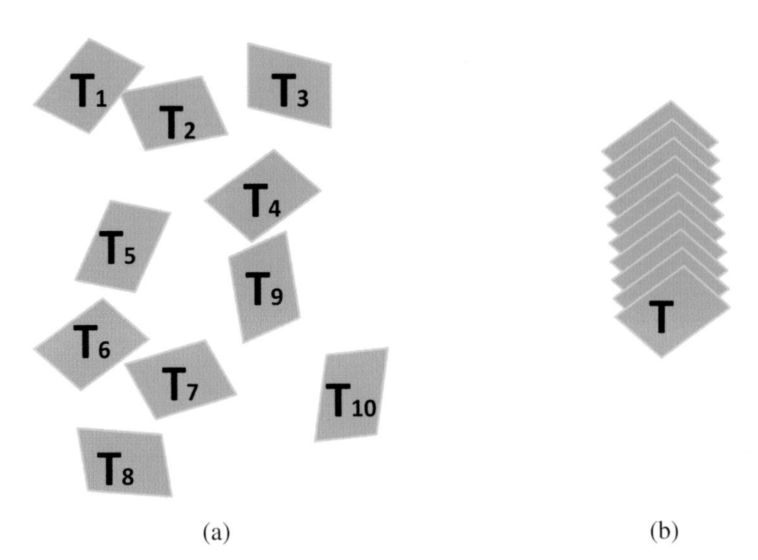


**Fig. 6.8.** The scattered and stacked collections of papers.

for which we can *calculate* exactly the entropies of two collections of well-defined systems — one ordered, the other disordered.

Consider a collection of boxes, each containing one mole of argon at a given temperature, and having volume $V$. For each box, we can calculate exactly its entropy [see the Sackur–Tetrode equation in Ben-Naim (2012)]. Now, I will let you prepare the collection of, say, ten boxes in any ordered or disordered arrangement you like. Three examples are shown in Figure 6.9. You can rearrange the order as you wish. The entropy of the collection of boxes in any of these arrangements (or in any other arrangement you wish) is exactly ten times the entropy of a single box of gas, *independently* of the *order* of the boxes.

Here is another untrue and misleading example. On page 30, we find:

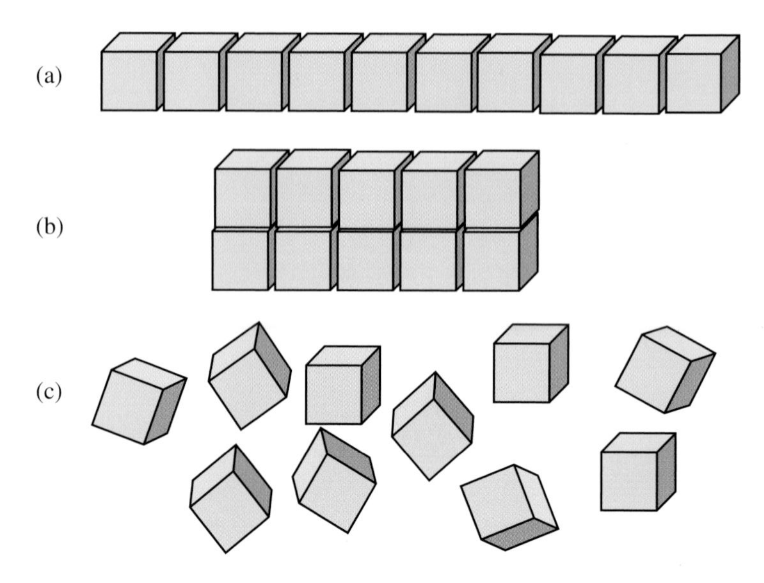


**Fig. 6.9.** Ordered and disordered ten systems.

> ...ice cubes in water melt but glasses of warm water don't spontaneously form ice cubes — share a common feature: Entropy *increases* throughout, as the system progresses from order to disorder. Whenever we *disturb* the universe, we tend to increase its entropy.

First, I do not know what is meant by "we disturb the universe." When we build houses, do we disturb the universe? When we make love, and a child is born, do we disturb the universe? Whatever your answer is and whatever you do, have no effect on the *entropy* of the universe, just as they have no effect on the *beauty* of the universe, the *wisdom* of the universe, the *stupidity* of the universe, or the *probability* of the universe.

Second, let us focus on the well-defined example given by the author. Figure 6.10 shows a glass of water with ice cubes.

**Fig. 6.10.** Three glasses of water and ice at different ambient temperatures.

It is not true that "ice cubes in water will melt." It depends on the ambient temperature. Ice at equilibrium with water at atmospheric pressure has a temperature of 0°. Now, if the glass is in the Sahara Desert, where the temperature is higher than 0°, the ice *will melt*, and the total entropy of the water in the glass will increase. On the other hand, if the glass is in Moscow in the winter, where the ambient temperature is −30°, all the water will turn into ice — and the entropy of the *water will decrease*. If the ambient temperature is exactly 0°, the ice cubes will *not melt*, and the total entropy of the water (in the two phases) will remain constant. The interested reader is referred to an example of a system whose entropy will oscillate during a whole year: Chapter 2, Ben-Naim (2015).

On page 32, we find other meaningless sentences:

> The beginning of our observable universe, the hot dense state known as the Big Bang, had a very low *entropy*. The influence

of that event orients us in time, just as the presence of the Earth orients us in space.

No one has the slightest idea what the entropy of the universe was at the Big Bang. In fact, to the best of my knowledge, *no one has ever defined* the entropy of the universe at the Big Bang. Obviously, the influence of that "event" on our orientation in time is as empty as the claim that that "event" influences our sexual orientation, my preference for sweet chocolate, or my wife's preference for dark chocolate. My orientation in time is not influenced by that "event." It is influenced by what my watch tells me!

On page 37, the author correctly describes Boltzmann's introduction of the entropy as a great leap forward in our understanding of entropy (not of the arrow of time). But then the author messes up the question of whether an isolated system will or will not return to its initial state, with the question of whether the entropy of the system will or will not increase or decrease. This mess is a by-product of the commonplace confusion over what the second law states. We have discussed this question in Chapter 5. Here, we briefly comment that if we remove a constraint in an isolated system — say, the partition in the expansion of gas, or in the mixing of two gases; see Figure 5.1b — the entropy change in these two processes will *always* — I repeat, will *always* — be positive (specifically, $R\ln 2$ in the expansion of one mole of an ideal gas, and $2R\ln 2$ for the mixing of two different ideal gases). The entropy of the system will *never* decrease — *never*, not extremely unlikely! On the other hand, we can ask whether the system will reverse to its initial state (i.e. all the gas molecules being in the left compartment, or the two gases separated in Figure 5.1b). For this question, the answer is yes.

The system can go to its final state. This event, however, is extremely unlikely to occur — not in billions of years, and not in billions of ages of the universe. Even in the rare event that this reversal will occur, the entropy of the gas (or the gases in the mixing process) will not change. (More on this in Chapter 5.)

The next section, "Entropy and Life" (pages 38–40), is in its entirety a sequence of sheer nonsense. I have commented on some of the statements in this section in Ben-Naim (2015), but I cannot resist the temptation of quoting the most beautiful sentence:

> Entropy, quite literally, makes life possible.

This statement is so absurd that I am willing to offer a substantial reward to anyone who will show me how entropy "makes life possible."

On page 40, the author asks a trivial question: "Why can't we remember the future?" Everyone, scientist and nonscientist alike, knows the answer to this question. We remember the past (more precisely, parts of our past) because the events are recorded in our brains, and if we are lucky we can retrieve this information from our memories. No events in the future are recorded in our brains, so there is nothing to remember from the future!

One can ask many similar silly questions, the answers to which are all trivial.

Why can't we remember the feelings of an insect?

Why can't we remember the taste of the apple Newton ate before he formulated his law of gravity?

Why can't we remember our mother-in-law's birthday?

The reader is urged to list some more events we do not remember, and then try to find the *deep* connection with the entropy at the Big Bang. Please send me a note if you find an interesting example, which I can include in the next edition of this book.

Perhaps the most absurd of all the absurdities is found on page 43:

> When it comes to the past, however, we have at our disposal both our knowledge of the current macroscopic state of the universe, *plus* the fact that the early universe began in a low-entropy state. That one extra bit of information, known simply as the "Past Hypothesis," gives us enormous leverage when it comes to reconstructing the past from the present.

This is probably the first time the idea of the Past Hypothesis appears in the book. It will appear again, and again, throughout the rest of the book. What does it mean? Nothing!

Besides, we *do not* have at our disposal the knowledge of the current state of the universe. We do not know the fact (fact?) that the early universe began in a low-entropy state. This is not a fact, but meaningless, fictitious nonsense! This "fact" and the corresponding Past Hypothesis appear so many times throughout the book that I felt I would soon start to worship the Past Hypothesis. And I forgot to mention the absurd and meaningless statement that the Past Hypothesis gives us *enormous leverage* when it comes to reconstructing the past from the present.

"To reconstruct the past from the present" — what a wonderful idea!

On page 50, the author discusses what happened before the Big Bang, and whether it is meaningful to ask such a question:

> So what happened before the Big Bang? Here is where many discussions of modern cosmology run off the trails. You will often read something like the following: "Before the Big Bang, time and space did not exist. The universe did not come into being at some moment in time, because time itself came into being. Asking what happened before the Big Bang is like asking what lies north of the North Pole.
>
> That all sounds very profound, and it might even be right. But it might not. The truth is, we just don't know.

At least the last sentence is an honest one. We are not sure about the event referred to as the Big Bang. We are not sure that time and space came into being at the Big Bang, and therefore we can legitimately ask what happened *before* the Big Bang, assuming that the Big Bang did occur.

Here are a few more meaningless statements:

> Our quest to understand the arrow of time, anchored in the low entropy of the early universe....(Page 51)
> So the relative smoothness of the early universe... reflects the very low entropy of those early times. (Page 54)
> ... if cosmologists had taken the need to explain the low entropy of the early universe seriously. (Page 55)
> We still don't know why the early universe had a low entropy... its entropy was small but not strictly zero.

And the rambling goes on, and on, and on. I would like to suggest a far more meaningful question to be added to the list: Why was the stupidity of the universe so low at the Big Bang, and why has it increased ever since? Knowing the answer to this question should have a profound effect on our life.

## *Chapter 8: Entropy and disorder*

As we have discussed in Chapter 5, the entropy of a system sometimes correlates with what we perceive as order or disorder. However, there is no general relationship — certainly not an identity between entropy and disorder. Figure 43 on page 153 is misleading, as it suggests that entropy both decreases and increases with time. Pure baloney!

Page 154 describes a system at equilibrium. Such a system does not change from the macroscopic point of view. For instance, its temperature, pressure, etc. are constant, or nearly constant, with time. However, the molecules in the system still jitter all the time.

To say that "such a system has no arrow of time" implies that some systems have an arrow of time, while others do not. This is very interesting. If the author will explain to me how one determines the arrow of time of each system, I promise to revise Chapter 1 of this book, and discuss all the arrows of time of all the systems in the universe.

The next chapter, on information and life, starts with a quotation from von Neumann when he suggested to Shannon to call his measure of information (see Chapter 1) "entropy," arguing that:

> No one knows what entropy really is, so in a debate you will always have the advantage.

This is, of course, not true. It is true that most people, including many scientists and many authors of popular-science books, do not know what entropy really is. But to claim that no one knows is an overstatement.

This chapter is full of nonsensical statements about information, entropy, and life. I have criticized some of them in Ben-Naim (2015).

On page 213, we find another misleading figure (54), showing that the entropy of a system fluctuates with time.

I have no idea from where the author took the "data" for such a graph. Was it calculated or measured for any real system?

The entropy of a system at equilibrium does not change with time. The configuration of the system changes, and Shannon's measure of information might change (see Chapter 5), but the entropy of the system is already the maximum value of the SMI, and this maximum value does not change with time!

Chapter 12 discusses black holes, the entropy of black holes, and the ends of time. I suggest to the author to compare *Brief* and *Briefer*. Perhaps this will convince him to omit this chapter in a future edition of the book (along with the rest of the book, of course).

Chapter 13, "The Life of the Universe," not only repeats endlessly the idea of the low entropy of the past of the universe, and the high (huge) entropy of the present universe, but also provides *numbers*! The author is careful to use the sign $\approx$ when he quotes some numbers for the entropy of the universe in the past, present, and future. He explains:

> The "≈" sign means "approximately equal to," as we want to emphasize that this is a rough estimate, not a rigorous calculation.

Doesn't the author know that by the very use of the sign ≈ he contributes to increasing the entropy of the universe? He should have said, more appropriately, that the sign ≈ as used in these pages of his book means *exactly* — not *approximately* — *meaningless* numbers.

On page 311, we find another "high entropy" absurd question and answer:

> Why don't we live in empty space?
>
> We began this chapter by asking what the universe should look like. It's not obvious that this is even a sensible question to ask, but if it is, a logical answer would be "it should look like it's in a high-entropy state...."

It seems to me that the author is relentless in producing *absurd* questions, then answering them with even more *absurd* answers. Thereby, he is creating consciously, or unconsciously, a *high-entropy* state of the reader's mind.

We don't live in an empty space, because there are no supermarkets where we can buy food in an empty space. And without food we cannot survive! And even if we could find food "in an empty space," what kind of life would it be without movies, shopping malls, and many other things that make life worth living?

Regarding the question "how the universe should look like," my "logical" answer is: It should look like a beautiful, low-entropy lady, and not an ugly "high-entropy state."

Here is another entertaining question. On page 314 the author asks:

> Why do we find ourselves in a universe evolving gradually from a state of incredibly low entropy, rather than being isolated creatures that recently fluctuated from the surrounding chaos? — does not yet have a clear solution. And it's worth emphasizing that this puzzle makes the arrow-of-time problem enormously more pressing.

Unlike the previous absurd answer given to an absurd question, here the author gives the *correct answer* ("does not yet have a clear solution") to a meaningless question. And, of course, the "arrow-of-time problem" is pressing. Indeed, the arrow of time is pressing, and therefore we should skip the next chapter (14) and move on to the last chapter, where we find more entertaining questions and answers. (Note the title of the chapter: "The Past through Tomorrow".)

> Why do we live in the temporal vicinity of an extremely low entropy state? (Page 339)
>
> The answer is trivial: We live in the *temporal vicinity* of a "low entropy state" because it is easy to reach that state by car, by bus, or even by walking. Only the very affluent people who own private planes can afford to live in the vicinity (sorry, temporal vicinity) of a "high entropy state."

The poor people will have no choice but to live in the vicinity of the bottom of the entropy. I wish I could help them in raising their entropy.

> Why does the Past Hypothesis hold within our observable patch of the universe? (Page 340)

The answer to this question is obvious. God *created* the Past Hypothesis, and that is why it holds in that patch of our universe, which is under His control. In other patches of the universe, He could not impose the Past Hypothesis! Simple as that!

On page 347, we are told about the "future hypothesis," and the possibility that the arrow of time will *reverse* at the moment the universe hits its maximum size....

This is really great. At that time it will be fun to live; the dead will rise from their graves, will "grow" younger, will walk backward, and will eventually return to their mothers' wombs. And, of course, the poor people in the last paragraph could afford to move to a higher-entropy state. What a refreshing idea.

Then we are shown through Figure 82 how the entropy first increases and then decreases with time. On page 352, in Figure 84, we see two more possibilities for the entropy to either increase monotonically or first decrease and then increase.

It is unfortunate that the author does not warn the reader that this is not science, or *science* fiction, but fictitious, incoherent ideas about entropy, time, and the universe.

In the epilogue, the author explains why he chose the title of his book. Perhaps he was right to choose (subconsciously) a meaningless title for a book full of meaningless discussions.

In a section titled "What is the Answer?" the author admits (page 367):

> Then after fourteen chapters of building up the problem, we devoted a scant single chapter to the possible solutions, and fell short of a full-throttle endorsement of any of them.
>
> We can state the problem very clearly but have a few vague ideas of what the answer might be.

In these words, the author admits that he has not provided a precise definition of time, as he promised on the first page of the prologue.

Even in the exceedingly long epilogue, the author repeats again, and again, the ideas about low entropy of the universe, etc., and even introduces a new meaningless idea that the idea of *time* itself is simply an "approximation." This idea is repeated four times on the same page (369) without explaining: "approximation" for what? Is time an approximation for time, for entropy, or perhaps for the universe?

Finally, after writing 375 pages of the book, the author concludes:

> Predicting the future isn't easy (curse the absence of a low-entropy future boundary condition!)... It's time we understood our place within eternity.

What a revealing conclusion. I always knew that predicting the future isn't easy (perhaps also impossible?). Now I know why! The culprit is the "low entropy boundary condition."

By the way, I fully understand our place within eternity. It is exactly three blocks from the lower-entropy intersection. Anyone who still does not understand, even after reading

Carroll's book, should write to me. I would be glad to explain that for a token fee.

Finally, I owe you an answer to the exercise I proposed on page 198. The entropy of the stacked papers (the "ordered" stack) is *higher* than that of the scattered collection (the disordered collection).

## 6.4 *Did Time Begin? Will Time End?* [Frampton (2010)]

This book poses two questions in its title. These questions are discussed in the preface and in the summary of the book. In between, a potpourri of topics are discussed which are irrelevant to the questions posed in the title of the book.

Needless to say, no answers are provided for these questions — another ho-hum book, as there is nothing new that we have not encountered in *Brief*, *Briefer*, and *Eternity* . Most of the preposterous ideas relating entropy, the second law, and time are repeated in this book.

The preface starts with the following two questions:

> Did time begin at a Big Bang? Will the present expansion of the universe last for a finite or infinite time?

Note that these two questions are deceptively similar, but very different from the questions posed in the title of the book.

Time could, or could not have, a beginning, independently of the occurrence, or nonoccurrence, of a Big Bang. Time could, or could not have, an end, independently of the present expansion (or future contraction) of the universe.

Then we find the following:

> The answers, which should become clearer in the next decade or two, could have *profound* implications for how we see our own role in the universe.

I strongly disagree with this statement. It only creates a false impression that the questions posed in the title of the book are very "important."

No one can tell when, and if, answers to these questions will ever be given in any foreseeable future. When answers will be available, they will have *zero*, not "profound," implications for how we see our own role in the universe." Such a fanciful but empty statement echoes the extraordinary ideas we have read in *Brief*, and in wholesale quantity in *Eternity*.

At the end of the preface, the author clearly — and I should also add, honestly — presents his aim for the book:

> This book aims to present some recently discovered scientific facts which can focus the reader's consideration of the two brief questions in the book title.

Thus, the author does not promise any *answers* to the questions posed in the title, but only that the recent discoveries "can focus the reader's consideration" of the two questions.

I read the entire book, and I did not see the relevance of most of the topics discussed in Chapters 1–7 to the questions posed in the title. I also failed to understand what the author means by focusing the "reader's consideration."

In fact, reading this book caused me to "unfocus" my considerations from the two questions. So many topics irrelevant to these questions are discussed in the book that only in the final chapter is the reader led back to these questions, and no answers are provided.

Most of the chapters 1–7 discuss some recent developments in cosmology, expansion of the universe, and the isotropy and homogeneity of the universe. Many concepts are mentioned without defining them — not even qualitative explanations such as "non-linear effects," "symmetry breaking," "violent infinities," and "quadratic divergence." A whole chapter is devoted to "dark matter" and "dark energy," and "negative pressure." All these render the book incomprehensible.

On page 55, we find one new outlandish concept which does not appear in *Brief* or *Eternity*:

> Photons which necessarily travel at the speed of light do exert a positive pressure and their equation of state is equal to plus one third.

I am not sure the lay reader knows what an equation of state is. I happen to know that an equation of state is an *equation* relating various thermodynamic variables. The best-known is the equation of state of an ideal gas, say argon. It is written as $PV = nRT$. Here, $P$ is the pressure, $V$ the volume, $n$ the number of moles of argon, $T$ the absolute temperature, and $R$ the gas constant. Clearly, an equation of state is an *equation*, not a number!

Here is another typical incomprehensible sentence (page 77):

> There is plenty of freedom in inventing a quintessence model and it is never completely clear whether the quintessence is any more than a parametrization of ignorance.

Reading this paragraph, I suddenly felt that my ignorance (of what is written in the book) became *totally parametrized.*

I admit that after reading Chapter 6 more than once, I have no idea what is written there. This chapter concludes with:

> Perhaps, the most striking mysteries are: What is the dark matter? What is the dark energy?

Of course, everything "dark" is a mystery. Why not include the most mysterious "dark entropy," and perhaps "dark time," then ask: Which is more mysterious — *dark* or *bright* entropy?

Chapter 7 introduces the entropy as a measure of disorder, which is the same baloney we have seen in *Brief* and *Eternity* (but not in *Briefer*!). Then, we are also given numbers: How big the (undefined!) entropy of the *present universe* is, and how low it was in the *early universe.* This is followed by relating the arrow of time to the second law and repeating all the nonsensical statements we have already encountered in *Brief* and *Eternity.*

Finally, the author admits:

> Reversal of the arrow of time is therefore an absurdity better to avoid if at all possible.

This comment should be addressed to those who write about "reversal of the arrow of time," not to the innocent lay reader, who has no idea what this means.

The rest of the chapter was totally unintelligible to me, and therefore I cannot comment on this. I will move to the last chapter, where the author brings back the questions posed in the title of the book.

As anyone can guess, there could be four possible answers to the questions posed in the title of the book:

1. Time had a beginning, and will have an end;
2. Time had a beginning, and will have no end;
3. Time had no beginning, but will have an end;
4. Time had no beginning, and will have no end;

No one can tell which answer is correct, and not even is it clear that one can assign probabilities to these events. Surprisingly, the author selects only three of four possibilities, ordering them according to their "*aesthetics*, in decreasing probability," with *conventional wisdom third*." (No. 2 on the list above.)

And how does the author "rate" the three possibilities (No. 3 is not considered by the author)?

On page 99, he says:

> Progress in addressing this question has been so rapid that it is possible to order these three futures, according only to their aesthetics.

Thus, there is no progress in *understanding* these questions, and no progress in *answering* these questions, only progress in *addressing these questions*! There is also much progress in the volume of writings about these questions.

I do not believe that rating the different cases according to their probability or likelihood is possible. I do not have the slightest idea how the author uses esthetics to order these cases! In my view, all these possibilities are equally, esthetically unpleasant!

Finally, the book contains a glossary of eight pages. Entropy, which the author claims is *central* to understanding the "arrow of time," does not feature in this glossary. Why? The reader who knows what entropy means can easily find the answer on pages 84–88 of the book. I have already discussed this in the previous sections of this chapter.

In conclusion, aside from the fact that there is nothing new in this book which is not found in *Brief*, *Briefer*, and *Eternity*, it has unfortunately resurrected some of topics related to the arrow of time to entropy and the second law, which were buried in *Briefer*.

# Epilogue

The main purpose of this book is not to tell you about the history of Time, but to critically examine what other writers have written about this topic.

As I have repeatedly said throughout the book, the most complete history of Time can be written in at most one or two pages. Time might have had a beginning ("birth"), and Time might come to an end ("death"). In between these two ends, nothing occurs to Time. In fact, even the occurrence of the two events — the beginning and the end — is extremely speculative. Personally, I doubt that they have any meaning in reality.

In the introductory part of this book, I asked you, the reader, to pause and think about a few questions. I posed questions that perhaps no one else has ever contemplated before — let alone provided answers to.

I hope that in reading this book you have been infected by my style of examining and criticizing well-known and well-recognized authorities. I hope I have contributed to honing your skill in reading carefully and critically anything you read, including this book.

If you are interested in assessing how much skill you have acquired, I suggest you read the quotations below, and after reading each one of them, answer the following questions:

1. Is the quotation meaningful?
2. If your answer is yes, write a short paragraph explaining its meaning to a layperson.
3. Once you are confident about its meaning, try to devise (or at least imagine) an experiment with the help of which you can verify the validity of the statement.

I would be glad to hear from you, and your opinions on these quotations, as well as any comments about this book.

*Quotations*

1. Entropy and disorder grow together. But nature also abounds in ordered structures.

   — *Sole and Goodwin (2000)*

2. The "arrow of time" is a metaphor invented by Sir Eddington to express the idea that there exists a purely physical distinction between past and future, independent of consciousness. Such a distinction is based on the *entropy principle*, which asserts that as time goes on energy tends to be transformed from an orderly into a less orderly form.

Time's arrow is irreversible, because entropy cannot decrease of its own accord without violating the second law of thermodynamics.

— *Campbell (1982)*

3. The ultimate source of order, of low entropy, must be the Big Bang itself.

— *Greene (2004)*

4. The past and future were different, and science could no longer ignore it. Thermodynamics gave science a wake-up call, forced it to grapple with the reality of linear time.

— *Schneider and Sagan (2005)*

5. Time, however, seemed to be like a model railway track. If it had a beginning, there would have been someone (i.e. God) to set the trains going.

— *Hawking and Mlodinow (2010)*

6. According to classical physics, time began at a moment when space was infinitely dense and occupied only a single point, and before that there were no moments.

— *Deutsch (1997)*

7. In a sense, our cells are eating energy, and their waste product is entropy.

— *Seife (2006)*

8. In the classical world, the arrow of time is associated with a steady increase of entropy, which can also be understood as a decrease of ordering.

— *Bruce (2004)*

9. According to nineteenth-century thermodynamics, closed systems gradually decline into disorder (their

entropy increases), and such a fate seemed to await the universe.

— *Smoot (1993)*

10. The statistical time concept that entropy = time's arrow has deep and fascinating implications. It therefore behooves us to try to understand entropy ever more deeply. Entropy not only explains the arrow of time, it also explains its existence; it is time.

— *Scully (2007)*[9]

# Appendix

**A sample of idioms involving time, and some illustrations:**

A stitch in time saves nine

Ahead of one's time

Bide time

Big time

Big-time spender

Crunch time

Devil of a time

Do time

Every time turns around

Give the time of day

Go down for the third time

Have a whale of a time

Have time to kill

Have time on your hands

High time

In the nick of time

It's feeding time at the zoo
Keep up with the times
Lost in the mists of time
Make up for lost time
Not give anyone the time of day
On the company's time
Pass the time of day
Pressed for time
Procrastination is the thief of time
Time is of the essence
Time to call it a day
Time to hit the road
Two-timer
When the time is ripe
Manage the clock
Take a timeout
Time and tide wait for no man
No time like the present
No time to lose

*Time runs*

*Killing Time*

*Time flies*

***Run after time***

***Pressed for time***

***Time hangs heavy***

*Time's birth*

***Sweet time***

***Time is relative***

***Time's up***

*Beware of the ravage of time*

*Time stopped*

*Behind time*

*Running out of time*

*Time heals*

*Time came to a halt*

*Saving time*

*Time is money*

*What a waste of time*

*Once upon a time, there was no time.*

*Time was born at some point in time, in a period of time when there was no time.*

*Time was born in a point of space, in a universe where there was no space.*

*A few minutes after its birth, Time grew apace.*

*One day, Time befriended Entropy, who was also born in a period of no-time, in a well-ordered universe of no-space.*

*Time and Entropy conspired to destroy, and disorder everything which stood in their way.*

*Today, we are witnessing the ravages of Time and Entropy in our midst.*

*Eminent scientists inform us that they have both good news, and bad news.*

*The good news is that Time will come to an end at some time, and will stop ravaging the universe.*

*The bad news is that Entropy will reach its maximum power, destroying the whole universe, including Time, before the end of Time.*

*I wonder which event will come first. Do you know the answer?*

# Notes

1. Regarding the nature of space–time continuum language, Tolman (1934) writes:

> In using this language it is important to guard against the fallacy of assuming that all directions in the hyper-space are equivalent, and of assuming that extension in time is of the same nature as extension in space merely because it may be convenient to think of them as plotted along perpendicular axes....
>
> That there must be a difference between the spatial and temporal axes in hyper-space is made evident, by contrasting the physical possibility of rotating a metre stick from an orientation where it measures distances in the $x$-direction to one where it measures distances in the $y$-direction, with the impossibility of rotating it into a direction where it would measure time intervals — in other words the impossibility of rotating a metre stick into a clock.

2. Euler's formula states that for any real number $x$ the following identity holds:

$$e^{ix} = \cos x + i \sin x.$$

Here, $e$ is the base of the natural logarithm, $i$ is defined as the square root of $(-1)$, i.e. $i = \sqrt{-1}$, and cos and sin are the trigonometric functions. If we substitute $x = \pi$ ($\pi$ is the ratio of the circle's circumference to its diameter, $\pi \approx 3.14159\ldots$), and since $\cos\pi = -1$ and $\sin\pi = 0$, we get the identity $e^{i\pi} + 1 = 0$.

3. The same is true of other results from the theory of relativity, such as the effect of speed or gravity on Time.
4. For the particular case of two compartments, the relationship is

$$\Pr(p, q) = \left(\frac{1}{2}\right)^N \frac{2^{[NH(p,q)]}}{\sqrt{2\pi Npq}}.$$

The more general result is

$$\boxed{\Pr(\{p_1, \ldots, p_n\}) = \left(\tfrac{1}{n}\right)^N \frac{2^{N \times H(\{p_1,\ldots,p_n\})}}{\sqrt{2\pi N^{(n-1)} \prod_{i=1}^{n} p_i}}}$$

For details, see Ben-Naim (2008, 2012).

5. If $v_x$, $v_y$, and $v_z$ are the *velocities* of a particle along the $x$, $y$, and $z$ axes respectively, the absolute velocity of the particle is defined by $v = \sqrt{v_x^2 + v_y^2 + v_z^2}$. This quantity is usually referred to as the *speed* of the particle. In this book we will refer to it as the velocity.
6. There have been many attempts to *derive* the second law of thermodynamics from the dynamics of the particles. Mackey (2003) devoted a whole book, *Time's Arrow: The Origins of Thermodynamic Behavior*, to this question. In fact, the first attempt to derive an equation for the "entropy" of a system which changes with time

and reaches a maximum at equilibrium was made by Boltzmann in his famous H-theorem.

In my opinion, the H-function defined by Boltzmann, and other functions based on the equations of motion of the particles, are not entropy functions but SMI functions. The latter can change with time and reach a maximum value at equilibrium. However, the *entropy* of the system is proportional to the maximal value of the SMI function. As such, the entropy of the system does not change with time.

7. Solution to Exercise 1: There are two solutions to the equation $p_4^2 + \frac{1}{16} = \frac{1}{8}$ . They are $p_4 = \frac{1}{4}$ and $p_4 = \frac{-1}{4}$. Clearly, you will not accept the negative solution for $p_4$. The only *real*, and meaningful, solution is $p_4 = \frac{1}{4}$.
Solution to Exercise 2: The equation to be solved is $p_4^3 + \frac{1}{64} = \frac{1}{32}$. There are three solutions to this equation (I got them by using Mathematica). They are

$$p_4 = \frac{1}{4}, \quad p_4 = \frac{-1}{8} - \frac{\sqrt{3}}{4}i, \quad p_4 = \frac{-1}{8} + \frac{\sqrt{3}}{4}i,$$

where $i = \sqrt{-1}$. Clearly, the only meaningful solution is $p_4 = \frac{1}{4}$. The other two solutions are not real numbers, and have no *real* meaning.

8. The entropy of the universe is not definable.

The entropy of a system is a *state* function. Giving a *state* of the system, say $(P, T, N_1, \ldots, N_c)$, one can define its entropy.

Regarding the universe, we do not know whether it is finite or infinite. We do not know all the kinds of particles in the universe, and how many there are of each. We do not know all the interactions between all the particles in the universe. Therefore, it is meaningless to talk about the entropy of the universe. It is *a fortiori* meaningless to talk about the entropy of the universe at the Big Bang, or at any other distant Time in the past. Thus, the so-called past hypothesis is a pure, meaningless hypothesis. Note also that the "entropy of the universe" cannot be determined either experimentally or calculated theory. For all these reasons, it is advisable to refrain from talking about the "entropy of the universe." Not in the present, not in the past, and not in the future.

9. Equating entropy with Time, or with the arrow of time, is not like equating apples with bananas (which are both fruits and quantifiable). It is like equating apples with psychology, or perhaps psychopathology (see also the dedication page).

# Bibliography


Albert, D. Z. (2000), *Time And Chance*, Harvard University Press, USA.

Atkins, P. (2007), *Four Laws That Drive the Universe*, Oxford University Press.

Barnett, L. (2005), *The Universe and Dr. Einstein*, Dover, New York, USA.

Ben-Naim, A. (2008a), *A Farewell to Entropy: Statistical Thermodynamics Based on Information*, World Scientific, Singapore.

Ben-Naim, A. (2008b), *Entropy Demystified: The Second Law Reduced to Plain Common Sense*, World Scientific, Singapore.

Ben-Naim, A. (2012), *Entropy and the Second Law: Interpretation and Misss-Interpretationsss*, World Scientific, Singapore.

Ben-Naim, A. (2015a), *Information, Entropy, Life and the Universe: What We Know and What We Do Not Know*, World Scientific, Singapore.

Ben-Naim, A. (2015b), *Discover Probability: How To Use It, How to Avoid Misusing It, and How It Affects Every Aspect of Your Life*, World Scientific, Singapore.

Brillouin, L. (1962), *Science and Information Theory*, Academic, New York, USA.

Bruce, C. (2004), *Schrödinger's Rabbits: The Many Worlds of Quantum*, Joseph Henry, Washington, DC, USA.

Callen, H. B. (1985), *Thermodynamics and an Introduction to Thermostatistics*, John Wiley & Sons, USA.

Campbell, J. (1982), *Grammatical Man: Information Entropy, Language, and Life*, Simon & Schuster, New York, USA.

Carroll, S. (2010), *From Eternity to Here: The Quest for the Ultimate Theory of Time*, Plume, USA.

Davis, P. C. W. (1974), *The Physics of Time Assymetry*, University of California Press, USA.

Deutsch, D. (1997), *The Fabric of Reality*, Penguin.

Eddington, A. (1928), *The Nature of the Physical World*, Cambridge University Press, Cambridge, UK.

Frampton, P. H. (2010), *Did Time Begin? Will Time End?* World Scientific, Singapore.

Greene, B. (1999), *The Elegant Universe. Superstrings, Hidden Dimensions, and the Quest for the Ultimate Theory*, Vintage, New York.

Greene, B. (2004), *The Fabric of the Cosmos: Space, Time and the Texture of Reality*, Alfred A. Knopf, USA.

Greene, B. (2011), *The Hidden Reality: Parallel Universes and the Deep Laws of the Cosmos*, Alfred A. Knopf, USA.

Hawking, S. (1988), *A Brief History of Time: From the Big Bang to Black Holes*, Bantam, New York, USA.

Hawking, S. and Mlodinow, L. (2005), *A Briefer History of Time*, Bantam Dell, New York, USA.

Hawking, S. and Mlodinow, L. (2010), *The Grand Design: New Answers to the Ultimate Questions of Life*, Bantam Books, London.

Einstein, A., Lorentz, H. A., Weyl, H. and Minkowski, H. (1952), *The Prince of Relativity*, Dover.

Mackey, M. C. (1992), *Time's Arrow: The Origins of Thermodynamic Behavior*, Dover, New York, USA.

Newton, R. G. (2000), *Thinking About Physics*, Princeton University Press, USA.

Penrose, R. (1989), *The Emperor's Mind: Concerning Computers, Minds and the Law of Physics*, Penguin, New York, USA.

Pross, A. (2012), *What Is Life? How Chemistry Becomes Biology*, Oxford University Press, USA.

Sainsbury, R. M. (2009), *Paradoxes*, Cambridge University Press, UK.

Schneider, E. D. and Sagan, D. (2005), *Into the Cool: Energy Flow, Thermodynamics and Life*, The University of Chicago Press, London.

Scully, R. J. (2007), *The Demon and the Quantum: From the Pythagorean Mystics to Maxwell's Demon and Quantum Mystery*, Wiley-VCH, Verlag GmbH & Co. KGaA.

Seife, C. (2006), *Decoding the Universe: How the Science of Information Is Explaining Everything in the Cosmos, from our Brains to Black Holes*, Penguin, USA.

Siegfried, T. (2000), *The Bit and the Pendulum: From Quantum Computing to M Theory — The New Physics of Information*, John Wiley & Sons, USA.

Smolin, L. (2004), *Time Reborn: From the Crisis in Physics to the Future of the Universe*, Penguin, UK.

Smoot, G. and Davidson, K. (1993), *Wrinkles in Time: Witness to the Birth of the Universe*, Harper Perennial, New York, USA.

Sole, R. and Goodwin, B. (2000), *Signs of Life: How Complexity Pervades Biology*, Basic, Perseus, New York, USA.

't Hooft, G. and Vandoren, S. (2014), *Time in Powers of Ten: Natural Phenomena and Their Timescales*, World Scientific, Singapore.

Tolman, R. C. (1934), *Relativity, Thermodynamics and Cosmology*, Oxford University Press, Oxford, UK.

Vedral, V. (2010), *Decoding Reality: The Universe as Quantum Information*, Oxford University Press, UK.

von Baeyer, H. C. (2003), *Information: The New Language of Science*, Harvard University Press, USA.

Weinberg, S. (1977), *The First Three Minutes: A Modern View of the Origin of the Universe*, Basic, New York, USA.

Weyl, H. (translated by H. L. Brose, 1950), *Space–Time–Matter*, Dover.

Woolfson, M. M. (2015), *Time and Age: Time Machines, Relativity and Fossils*, Imperial College Press, UK.

# Index

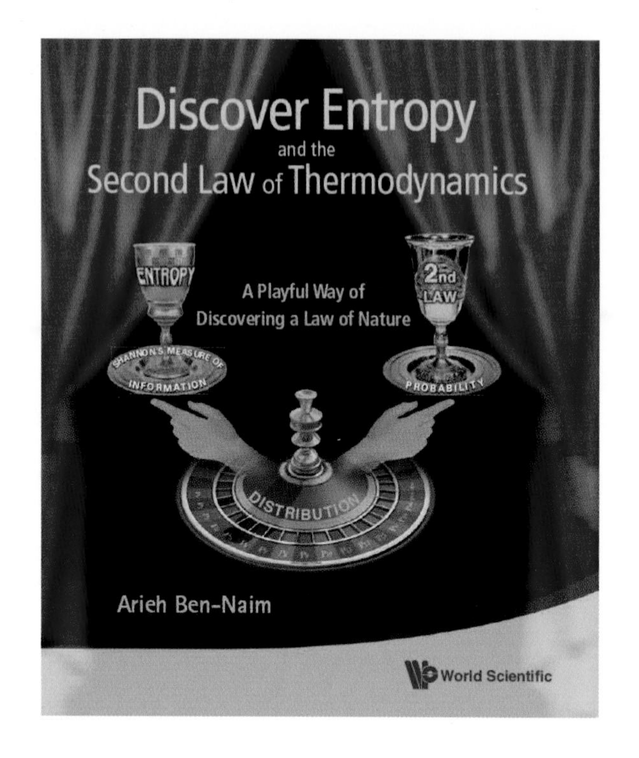
Discover Entropy
and the
Second Law of Thermodynamics
ENTROPY
A Playful Way of
Discovering a Law of Nature
2nd
LAW
SHANNON'S MEASURE OF
INFORMATION
PROBABILITY
DISTRIBUTION
Arieh Ben-Naim
World Scientific